PRAISE FOR *TEAMING WITH MICROBES* BY JEFF LOWENFELS

"Digs into soil in a most enlightening and entertaining way."
—*Dallas Morning News*

"Should be required reading for everyone lucky enough to own a piece of land."
—*Seattle Post-Intelligencer*

"Required reading for all serious gardeners."
—*Miami Herald*

"One heck of a good book."
—*The Oregonian*

"A very helpful source."
—*Sacramento Bee*

"A must read for any gardener looking to create a sustainable, healthy garden without chemicals."
—*Detroit News*

"A breakthrough book for the field of organic gardening . . . . No comprehensive horticultural library should be without it."
—*American Gardener*

"This intense little book may well change the way you garden."
— *St. Louis Post-Dispatch*

"Sure to gain that well-thumbed look that any good garden book acquires as it is referred to repeatedly over the years."
—*Pacific Horticulture*

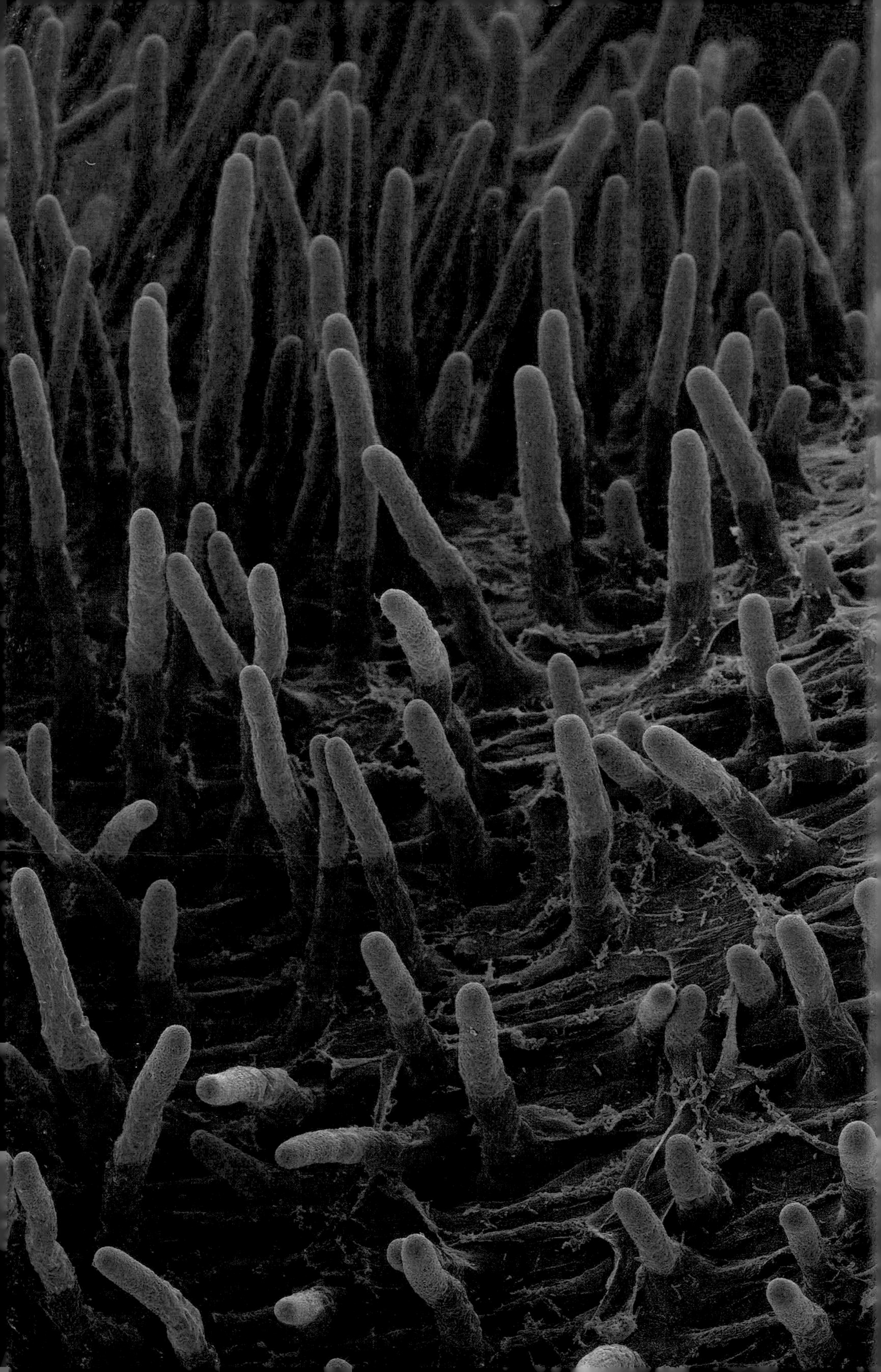

TEAMING WITH NUTRIENTS

The Organic Gardener's Guide to Optimizing Plant Nutrition

JEFF LOWENFELS

TIMBER PRESS
Portland · London

Frontispiece: Scanning electron micrograph of root hairs developing on a radish during seed germination

Photo and illustration credits appear on page 250.

Published in 2013 by Timber Press, Inc.

The Haseltine Building
133 S.W. Second Avenue, Suite 450
Portland, Oregon 97204-3527
timberpress.com

2 The Quadrant
135 Salusbury Road
London NW6 6RJ
timberpress.co.uk

Printed in China
Book design by Susan Applegate
Composition and layout by Holly McGuire

Library of Congress Cataloging-in-Publication Data
Lowenfels, Jeff.
Teaming with nutrients: the organic gardener's guide to optimizing plant nutrition/Jeff Lowenfels.—1st ed.
p. cm.
Includes index.
ISBN 978-1-60469-314-0
1. Plants—Nutrition. 2. Plant nutrients. 3. Organic fertilizers. I. Title. II. Title: Organic gardener's guide to optimizing plant nutrition.
QK867.L87 2013
575.7′6—dc23 2012033926

A catalog record for this book is also available from the British Library.

CONTENTS

FOREWORD

UNLESS YOU'RE a scientist who deals with mycorrhizae, you've probably never given much thought to how plants eat. Most gardeners think that growing a good tomato is all about photosynthesis and mixing in some nitrogen, phosphorus, and potassium (N–P–K). Jeff Lowenfels shows how wrong this assumption is.

This new book is the perfect companion to Jeff's first book, *Teaming with Microbes: The Organic Gardener's Guide to the Soil Food Web,* which deftly explains the all-important workings of the soil food web in delivering nutrients to plants. Now you can learn what plants do with these nutrients, how they get them inside their roots, and what happens to nutrients once they are in the plant.

I've studied the science of how plants take in nutrients for most of my career. It's taken a lot of chemistry and biology to get me to where I am today. I only wish I'd had this book as a much younger man just starting out in plant science and mycology—it would have saved me a lot of time and painful learning. Jeff has the knack of being able to explain complicated science in ways that are instantly understandable and even enjoyable. He holds your hand in the rough spots, walks you through the science, and then takes off once he knows you've grasped the concepts.

Jeff's book is as timely as it is informative. Too many gardeners think they are taking the modern path by blindly pouring on synthetic N–P–K fertilizer in accordance with a picture on the label or an ad on television. We let chemistry take over. We know little about what we're doing, but we do it anyway. The result has been an alarming spike in phosphorus and nitrogen pollution.

N–P–K gardeners owe their practice to the great scientist Justus Von Liebig, the father of artificial fertilizers. What most don't know is that later in his life Von Liebig acknowledged a grave mistake in relying only on chemistry. In fact, he saw the negative impacts of artificial fertilizers on life in the soil in his own vegetable garden and thereafter preferred organic matter to the inorganic chemical fertilizers he invented. To quote Von Liebig, "After I learned the reason why my fertilizers weren't effective in the proper way, I was like a person that received a new life." Understanding how nutrients work will make us all see the light and ultimately make us better gardeners with more sustainable gardens.

I work with gardeners around the world. No matter their language, culture, or age, the story is always the same: all gardeners want to be successful. The good ones are those who learn as much they can about how plants function because they know this information will allow them to counter what they can't control. I am quite sure the information in this book will greatly add to any gardener's knowledge base and allow him or her to more perfectly balance those factors that can't be controlled.

How plants eat—it's about time we all learned.

—Dr. Mike Amaranthus

Chief Scientist and President, Mycorrhizal Applications, Inc.

ACKNOWLEDGMENTS

THERE ARE TWO people who must be acknowledged. The first is Judith Hoersting, my best friend and wife, who shared with me cellular biology, chemistry, essential nutrients, my MacBook, and the Portland library for a year while I got this out of my system. The second is Lisa D. Brousseau, who appeared out of nowhere and edited this book into shape. She kept the story on track; provided useful insights, corrections, and suggestions; and suffered for having to work with the worst proofreader in the history of the written word.

I sometimes think that never blows so red
The Rose as where some buried Caesar bled;
That every Hyacinth the Garden wears
Dropt in her Lap from some once lovely Head.

Omar Khayyam

Eleventh-century Persian astronomer-poet

INTRODUCTION

HOW DO plants eat? I am pretty sure this is an age-old question. It probably came up 10,000 years ago after some early gardener noticed that rotting fish did wonders for plants. The observation that one's urine had a beneficial impact on plants could not be missed, either. These and other natural fertilizers not mentioned in public helped trigger the Neolithic Revolution, the transition from hunter-gatherer to farmer-gardener. Even in ancient times, feeding an ever-growing population required horticultural advances. The Aztec and Mayan civilizations, for example, were all about growing food to support burgeoning populations. They offered their gods sacrificial blood to ensure a good harvest. Perhaps this practice arose from their observation that soil bloodied from butchering an animal or as a result of some mortal blow during a heated battle grew better plants.

I come from a long line of natural fertilizer users. My grandfather and dad taught me to bury the uneatable, bony fish we used to catch every summer under roses and tomatoes. We had a horse for a while, too, and chickens, geese, ducks, and rabbits. We knew about the wonders of manure. I won't go into my use of urine as a fertilizer, but with three boys growing up on eight acres, you can bet it was applied to all manner of plants liberally, with varying impacts on the plants. Today, gardeners use homemade and commercial fertilizers, composts, and mulches. Many simply follow the directions with little or no thought about what those powders and liquids really do. We're just glad they do it.

After more than 50 years of gardening, I realized that I didn't know much about fertilizers other than what I picked up from my family and my own observation over the years. When I started to ask my gardening friends what they could tell me about fertilizers, I discovered a startling fact: I couldn't find one gardener who could tell me how they work. It seems that today's gardeners are just as clueless about how fertilizers work as were our early ancestors. We rely on the same principle, observation, which, sad to say, includes advertising.

Still, how plants eat has been the subject of discussion probably ever since the early days. Ancient Greeks, for example, engaged in arguments about what should be used as fertilizers, and these sound like modern organic versus inorganic arguments. The seventeenth-century scientist Jean Baptiste van Helmont proved that plants actually didn't need fertilizers, only water. A century later, the Englishman Jethro Tull and others promoted the notion that the carbon in plants came from organic particles in the soil that only entered through the roots. It is from here that Tull came up with the idea of pulverizing soil particles to make them more edible to the roots, and the practice of rototilling was born.

In the mid-1800s, the German chemist Justus Von Liebig showed that plants actually get carbon from the carbon dioxide in the atmosphere. He did so, incidentally, by demonstrating that after plants die and decay, they leave soil richer in carbon. He also examined the ashes of burned plants and saw they contained minerals that obviously didn't come from organic humus or soil particles. From these studies came Von Liebig's Mineral Theory of Plant Nutrition, which states that plants need certain minerals and that these minerals can be put back into the soil in inorganic forms to enable plants to grow in soils depleted of these minerals. This was pretty heady stuff in its day—and it was controversial.

Von Liebig experimented in a field in the English countryside from 1845 to 1849, growing crops using artificial manures, the first synthetic fertilizers. He made one mistake, however. Von Liebig thought plants got their nitrogen from the atmosphere, so he didn't add it to his soils. His plants didn't do well. Quickly proven wrong on where nitrogen came from, he and others prepared inorganic fertilizer formulations that contained nitrogen and worked well. In fact, they worked as well as manures. After only a few years of testing, the results were so impressive that they caused Von Liebig to predict, "A time will come, when fields will be manured with a solution of glass (silicate of potash), with the ashes of burnt straw, and with the salts of phosphoric acid, prepared in chemical manufactories, exactly as at present medicines are given for fever and goitre."

Von Liebig's experiments and the efforts of competing contemporaries led to the birth and growth of the chemical fertilizer industry. The first fertilizers of this kind were sold in 1845, and thus began a huge shift in how society lived. For the first time in history, farmers did not have to be dependent on animals for manures. They could spread inorganic chemicals on their fields instead, chemicals that could be made in factories or mined from the earth.

The importance of this change can't be missed by anyone who has had just a dog or cat. Animals require costly feeding and lots of time-consuming care. They take up valuable land, and consume a lot of nutrients. The invention of artificial manures meant that one could farm without having any animals (though I would hope the dog, at least, could stay).

Before the 1850s, when a farmer exhausted his land he either had to resort to using expensive, animal-based manures or, as American farmers did, find fresh land. The American West was settled in part because early American farmers didn't have enough animals to produce enough manure to replenish their soils. The American slogan that expansion of the country was "manifest destiny" had more to do with farming than we're taught in school. When the nutrients in the soil ran out, it was cheaper and easier to just move west and start over.

Few farmers really cared, much less understood, why these inorganic chemicals worked, just as long as they did so. What else was there to know? However, the new science of agriculture was burgeoning. Justus Von Liebig, considered by most as the father of modern agricultural science, again contributed to the field in 1863 when he proposed the Law of the Minimum. It states, "A manure containing several ingredients acts in this wise: The effect of all of them in the soil accommodates itself to that one among them which, in comparison to the wants of the plant, is present in the smallest quantity." In other words, plant growth is limited by the least abundant mineral, no matter how abundant the other minerals happen to be. This idea affects how fertilizers are used even today. (Incidentally, Von Liebig may have appropriated the Law of the Minimum from Carl Sprinkle, who at least deserves mention here.)

The application of the Law of the Minimum revealed that the

elements nitrogen (N), phosphorus (P), and potassium (K) were used by plants in the greatest quantities and were the nutrients plants most needed if they were to thrive in depleted soils. This resulted in the era of measuring the amount of nitrogen, phosphorus, and potassium (and other essential plant nutrient elements) and amending deficiencies with inorganic fertilizers. Today, any fertilizer sold around the world carries three numbers on the package representing the percentage by weight of each of those elements: the N–P–K ratio.

Consequently, many gardeners think they only need to look for these three numbers to know which fertilizer to use. In fact, fertilizer manufacturers are making it even easier. These days all a gardener needs to do is look at the representative pictures on fertilizer packages. Spot the tomatoes, annuals, perennials, houseplants, or orchids you want to feed and buy that package. Of course, there is all manner of advertising dollars spent to convince you that you need to know even less: just buy a particular brand, and your plants will do fine.

Fortunately, like Neanderthals, those early Neolithic gardeners, most of us are able to make a bit of sense of things as a result of our own observations. Gardeners today know that if you put lots of nitrogen on the lawn, it grows like crazy and has to be mowed more often. So nitrogen has something to do with leaves and growth. Yellow leaves with green veins turn back to green leaves with green veins when iron is added to the soil. So iron must have something to do with the plant being green.

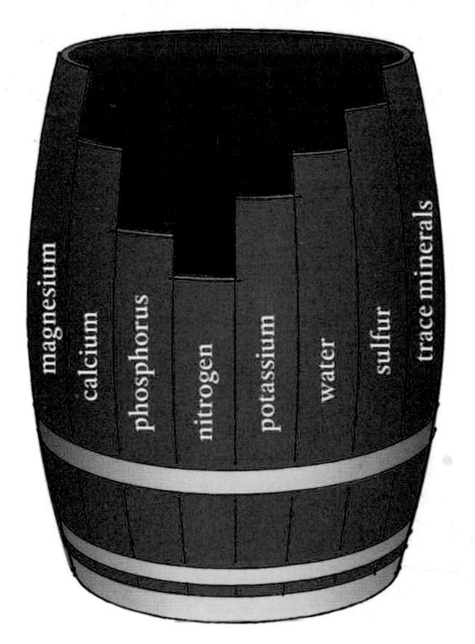

Von Liebig's Law of the Minimum states that if one nutrient doesn't reach the minimum required by a plant, then it doesn't matter how much more of the others you apply. In short, the barrel will only hold up to its lowest stave.

Still, many gardeners simply believe in the power of fertilizers because they are exposed to advertising. Phosphate, we have been told, helps grow healthy roots. Unless you dig up roots, who can really tell? But I am a believer. Boron is needed for pollination. O.K. And calcium or magnesium has something to do with preventing spots on tomatoes.

WHY DOES IT MATTER?

Although gardening is all about growing plants, many gardeners don't understand how nutrients actually get inside plants so they can grow in the first place, or how nutrients contribute to growth (and better flowers, tastier vegetables, healthier trees) once inside. By understanding how fertilizers work, how they get into plants, and what they do thereafter, you won't have to rely on someone else who is only guessing what your plants need. You will know something about how fertilizers work and whether what you are paying for is worth it. Information is power.

Knowing how fertilizers work should also make you a more sustainable gardener and help the planet. Take nitrogen, for example. The production of nitrogen-based synthetic fertilizers is a very energy intensive manufacturing process. More than 5 percent of the world's natural gas production is used to make these fertilizers. Less than 100 years ago, all of the plant available nitrogen in the world, except for a tiny bit fixed during electrical storms, was produced by microbes. All of the nitrogen in your body was naturally produced, whereas today half of your nitrogen comes from synthetic sources. In addition, the excessive use of nitrogen fertilizers causes severe pollution. Gardeners use three times the nitrogen per acre that farmers use. This results in excess nutrients that are washed into waterways and harm aquatic ecosystems.

Using plant-essential nutrients the proper way will help head off some serious environmental disasters already in the making. Manufacturing nitrogen so that gardeners can waste it in prodigious amounts is not our only problem. The world is about to reach the peak of its ability to produce phosphorus. The precipitous decline in availability of this key nutrient has led some to predict that there is less than 50 years' worth remaining. Gardeners had better use sustainable practices, or we will really test the Law of the Minimum with regard to essential plant nutrients. This is important stuff.

Researchers are studying the uptake and use of nutrients by plants because it influences many important technologies that affect us all. The genetic modification of plants and the creation of organic and inorganic pesticides, herbicides, and biocides all have to do with how plants eat. For example, many herbicides work by interfering with a plant's nutrient system, preventing some nutrient from getting into the plant and causing it to die. Many genetically modified plants are engineered to resist glyphosate (available to gardeners as Roundup). As more genetically modified crops are being grown, an increasing amount of glyphosate is making its way into the natural environment. If you don't understand how the plant nutrient system works on a chemical level, you can't understand the current GMO-glyphosate debate. Once you understand how plants eat, you'll realize where the debate is going to end, and you can make some decisions about whether to use glyphosate in your yard.

Plants are what they eat. If you want the best possible plant, you need to feed it properly. It helps to know what each essential element really does and the best sources for them. Along the way, I hope you become infected by the knowledge of the wondrous universe inside plants: cells, organelles, secret pathways, manufacturing on a scale of unimaginable proportions—life at its most basic, simplest form. Seeing the beauty in this system will surely enhance what you already feel about plants and gardening. I found myself comparing cellular systems to human systems to planetary systems and ended up confronting the meaning of life, all while zeroing in on the plant cell, its structure, and wondering (in the truest sense of that word) about the amazing workings. Plant cells build everything that plants are, which results in much of what we are. It's truly awesome.

THE SOIL FOOD WEB

While this is a book about how plants eat and includes some suggestions about what kinds of fertilizers to use on plants, let me assure my loyal readers that I am not abandoning the principles of the soil food web outlined in my previous book, *Teaming with Microbes: The Organic Gardener's Guide to the Soil Food Web*. I now have a better understanding of the importance of mycorrhizal and actinorhizal relationships with plants. It is important to establish and maintain these and other soil organism so that they, in turn, can feed your plants. The practices recommended in this book are organic and are not chemically oriented. All of the nutrients plants require can be easily supplied without having to resort to any of Von Liebig's artificial manures.

Von Liebig eventually recognized the negative impacts of artificial fertilizers on life in the soil of his own vegetable garden. Thereafter, he preferred the use of organic matter to inorganic fertilizers. In fact, he spent a good part his later life arguing that the Britons should use their sewage as fertilizer, this despite having created the inorganic fertilizer industry.

A ROADMAP TO THE BOOK

Describing the process of how plants take in and use nutrients necessarily involves chemistry and biology. Don't worry. Each chapter builds on the previous one, so by the time you get to the punch line, you will have the science under your belt and get it. Let's take some of the mystery out of it right now with a short summary.

In chapter 1, I discuss the various parts of a typical plant cell, because this is where the action takes place. The outer cell wall and the plasma membrane up against it act as barriers and regulators of what can enter and leave a cell. Special membrane proteins assist water and nutrients in entering the cell, while keeping unwanted materials out. The cytoplasm holds structures and organelles that perform special jobs related to taking up and using fertilizers. They provide power to the cell and serve as sites for photosynthesis. The nucleus is the command center where the DNA is housed. Cells have transportation and communication infrastructure, protein construction areas, and even tunnels that connect every single cell in a plant.

Chapter 2 covers the necessary basic chemistry needed to understand the journey of nutrients. You don't have to remember anything from school. I discuss atoms, electrons, and chemical bonds. (Finally, we have a reason to know about covalent, ionic, and hydrogen bonds, which affect the qualities and availability of the various nutrients.) This chemistry results in the four types of molecules that are necessary for life: carbohydrates, proteins, lipids, and nucleic acids. I also describe ATP (adenosine triphosphate), the energy currency in all cells, and enzymes that speed up the millions of chemical reactions that occur within plant cells.

In chapter 3, I discuss the botany that affects nutrient uptake and utilization. Four kinds of plant tissues and their organization into special organs (leaves, stems, and roots) have roles in the uptake of nutrients. Some play surprising and unexpected roles, including aiding in the formation of symbiotic relationships and other biological partnerships important to nutrient uptake.

The seventeen elements essential for the lives of plants are covered in chapter 4, including the macronutrients (nitrogen, phosphorus, potassium, calcium, magnesium, and sulfur) needed by plants in the greatest quantities and the micronutrients (boron, chlorine, copper, iron, manganese, zinc, molybdenum, and nickel), which are needed in only trace amounts. The other three essential nutrients, carbon, hydrogen, and oxygen, are also covered, as are the different degrees of mobility of the various nutrients and its implications.

Water plays the starring role in this story. In chapter 5, the chemistry and botany described in the previous chapters are used to help explain how water moves through the soil to get to and then into the roots of a plant. Once water is inside a plant, there are different ways it can to get to the xylem, where it is carried up to the stems and leaves along with the nutrients dissolved in it. Water is later moved throughout the plant in the phloem, along with the sugars, proteins, enzymes, and hormones produced in plant cells. I describe the interplay between the two tissues in the vascular system when it comes to plants taking up and distributing nutrients.

Chapter 6 covers the movement of nutrients into and then inside a plant, starting with their movement in the soil around plant roots. Once inside the plant, nutrients must be transported across cell membranes so they can be used to make all the compounds the plant needs for growth and maintenance.

In chapter 7, I explain the role of the essential nutrients in the makeup of the four molecules of life. Carbohydrates are produced via photosynthesis. Proteins are constructed from various combinations of the twenty amino acids. Lipids are made up of fatty acids and glycerol. Finally, nucleic acids are the molecules DNA and RNA that carry the genetic code.

In chapter 8, the book gets down to actual gardening and applying some of the knowledge gained in the previous chapters to make our art more of a science. I discuss whether you even need to fertilize, and, if you do, what steps you should take. The use of fertilizers should be based on sound knowledge, which can only be obtained by having your soil tested.

In chapter 9, I discuss the other factors that influence nutrient uptake and the use of nutrients: temperature, soil microbes, moisture, soil compaction, and the chemical reactions that occur within plants and in soil. Many gardening practices come into focus when the science behind them becomes clearer.

Finally, in chapter 10 I offer recommendations of what to feed plants based upon knowledge of how plants take up nutrients and how they use them. I provide fertilizer recipes designed for annuals, vegetables, and lawns and describe the best ways to apply fertilizers, including the timing of application, and other characteristics of commonly obtainable natural fertilizers.

A FEW FINAL THOUGHTS

I am the first to admit that to understand how plants take up nutrients requires more botany, chemistry, and cellular biology than the average (and probably above-average) gardener knows. This doesn't mean you have to learn college-level science before getting it. Also, we are gardeners, not chemists, and it is my book. So, in my book I avoid chemical equations like the plague. I don't need to explain everything. That's what the Internet is for. I supply enough information so you can get to the end of the story without being overwhelmed—or at least to know where to start using the Internet.

And for goodness sake, don't read this book like it's a textbook. There is a lot here, sure, but if you just read it and don't get hung up on memorization, you'll find that pretty soon the lingo flows and the understanding increases with each chapter. It helps to look at the illustrations, which were drawn especially to depict things in three dimensions. Everything builds on what came before, and all will be clear. Just relax and read for fun.

So, let's go. Hopefully, you will never again take for granted what happens to make your plants grow.

ONE The Plant Cell

WHAT'S IN THIS CHAPTER

- **The outer cell wall** and the plasma membrane are barriers that regulate what can enter a cell.
- **The cell membrane** has tunnels that connect all the cells in a plant.
- **Aquaporins** and transport proteins embedded in the membrane assist water and nutrient ions in entering the cell, while keeping unwanted materials out.
- **The cytoplasm** is the liquid part of the cell along with the organelles, the structures in a cell that perform special jobs.
- **Mitochondria** are the powerhouses of the cell.
- **Chloroplasts** are the sites of photosynthesis.
- **Protein molecules** are constructed on ribosomes located on the endoplasmic reticulum or floating in the cytoplasm.
- **The central storage facility,** the vacuole, is surrounded by another membrane, the tonoplast.

- **Lysosomes and peroxisomes** are the recycling centers of the cell.
- **The nucleus** is the command center where the DNA is housed.
- **Microtubules** and actin filaments serve as the communication centers and transportation infrastructures.

THERE ARE LOTS of places we could start. However, I am a big believer in setting up the stage before introducing the players. When it comes to how plants eat, the stage is the basic plant cell. The Cell Theory holds that the cell is the smallest unit of life and that all living things are made up of one or more cells. Moreover, all cells come from pre-existing cells.

Most gardeners studied some cellular biology in school, and many of us have a basic understanding of how cells divide. In recent years, however, scientists have made many new discoveries by studying DNA and gene markers and using bigger and more sophisticated electron microscopes and a whole host of other wonders of modern technology. These discoveries have resulted in a much clearer picture of the plant cell and what goes on outside and inside it.

SIZE

Plant cells are incomprehensibly small and thus outside the purview of a gardener's day-to-day thinking. It probably doesn't help to note that most plant cells are in the range of 10 to 100 micrometers (0.0003937 to 0.003937 inches). Micrometers (or microns) and nanometers (there are 1000 nanometers in a micron) are the measurements used at the cellular level. However, it is helpful to at least try and put the size of a plant cell into some perspective.

First, let's forget about inches from here on, except to note that 1 meter is about 39 inches. A centimeter is one-hundredth of a meter and getting a lot closer to cell size. It is just smaller than the diameter of an American dime. A millimeter is one-thousandth of a meter and is the thickness of that dime. A micron is one-millionth of a meter and is the thickness of a cell.

Cell components are measured in nanometers, or billionths of a meter. Nanometers are used to measure atoms and molecules. Despite all of these numbers, these small sizes are nearly incomprehensible. Let's try. A human hair is about 100,000 nanometers wide. You can see one without a microscope and unaided with the naked eye. A water molecule is less than 1 nanometer wide. You can't see that with a microscope in a high school biology class. Anything smaller than 500 nanometers can only be seen by using an electron microscope.

Still not getting how ridiculously small this is? If each person on Earth were 1 nanometer tall, we would all fit into a commercial seed packet. That is more than 6.5 billion very tiny people, by the way. Or consider the period at the end of this sentence; it is about 500 to 600 microns in diameter. Five to fifty plant cells would fit into this area. Don't forget they are three-dimensional.

Another way to appreciate these sizes is to pretend that a single atom is the size of a small pea. A pea pod, then, is the size of a simple molecule. A semi-trailer truck full of pea pods is the size of a complex molecule, and an ocean-going transport tanker that carries a whole bunch of pod-filled semi-trailers is the size of a cell.

Size may seem like a limiting factor when it comes to cells. One might think that for efficiency and economy cells should be large. However, a smaller sphere has relatively more surface area per unit inside area. If a cell were to get too large, there would not be enough membrane surface to supply the cell with sufficient nutrients to survive. In any case, there is a practical limit on the size of cells, and small is not a bad thing.

The number of cells in a plant is equally stunning. A 20-year-old birch tree is made up of something like 15 or 16 trillion cells. Probably more, but who can count that high? Again, these plant cells are not flat objects, as they are so often depicted in school science texts. That tree is decidedly three-dimensional, as are every one of its trillions of component cells. They have thickness and are filled with equally three-dimensional parts composed of three-dimensional atoms, the smallest particle of an element.

Everything from here on should be viewed in three dimensions. By the end of this chapter, you should able to mentally shrink down to a half a micron and wander around a cell, inspecting its parts and getting a molecule's eye view of things. At that size, our plant cell looks like a factory as large as a football stadium.

THE CELL WALL

Plant cells are surrounded by a cell wall, a strong, lattice-like structure. The cell wall is what distinguishes plant from animal cells, which lack cell walls. Averaging around 20 nanometers thick, its primary function is to contain and protect the cell.

The cell wall is mostly made up of cellulose, which is what makes the fibers of the pages of this book (unless it is electronic). About one-third of all vegetative matter is composed of cellulose, making it the

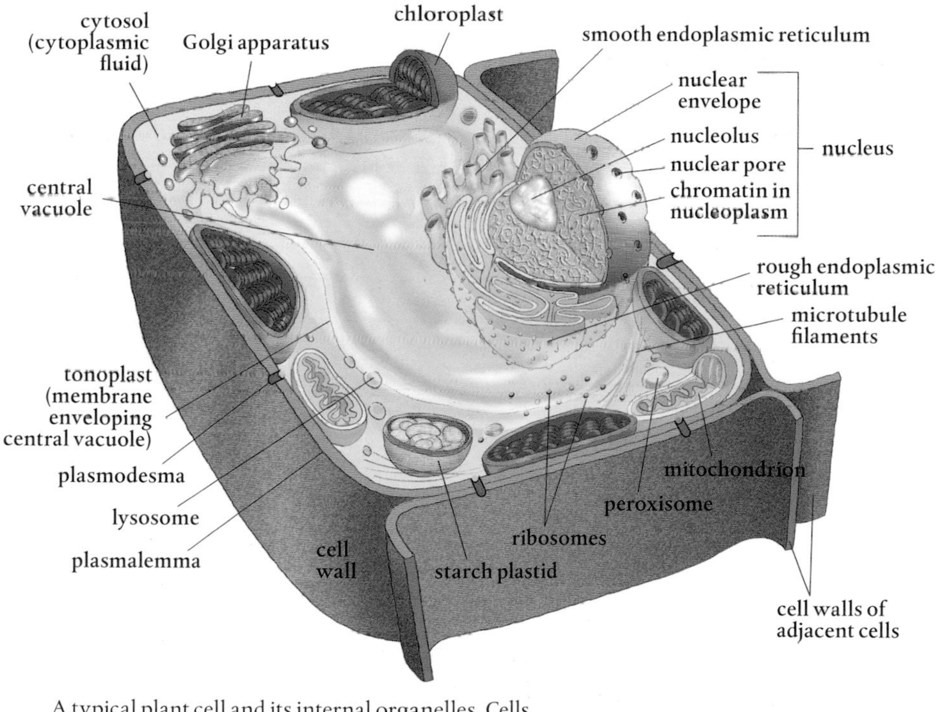

A typical plant cell and its internal organelles. Cells may be small, but they are still three-dimensional and should always be visualized in that form.

most common biologically produced substance on Earth. Cellulose is a complex molecule made up of many simpler glucose molecules, a type of sugar, that when linked together become polysaccharides (meaning "many sugars"). The structure of polysaccharides allows them to be woven closely to form long strands. These fibrils overlap each other to create layers, which gives plant cell walls great strength. Lest the image of a solid wall confuse things, the cell wall is actually extremely porous. Imagine the cell wall as a luffa sponge.

Plants are not the only organisms to produce cellulose. Some bacteria are able to make it as well. This fact gave rise to the concept that bacteria were ingested and eventually incorporated into plant cells with great success, for the bacteria and the cells, the result of which is that all plant cells now produce cellulose.

Special protein complexes produce the cellulose fibers of the cell wall. These proteins move along the outer membrane of the cell, which lies just inside the cell wall. These proteins spool out fibrils almost like the spindle on a knitting machine or a cotton candy machine. The supplies that make up the cell wall (including cellulose, hemicelluloses, and pectins) are all produced inside the plant using nutrients (such as boron) that are aggregated there from the soil. The nutrients are transported to the cell membrane for the construction of the cell wall on the other side

This scanning electron micrograph shows cellulose fibers, which give great strength to the cell wall.

of the membrane. This is quite an engineering feat, one of many accomplished by these tiny cells.

Consider that there are about 3000 molecular units linked together in just one of the polysaccharide chains that form the microfibrils that are woven into cellulose and cell walls. The synthesis of cellulose in the cells requires enzymes, nitrogen-based molecules that vastly speed up the chemical reactions that occur in a cell. If gardeners had to wait for the right sugar molecules to come along and collide with another 3000 times to make just one strand, there is no way we would live to see a plant grow.

Cell walls are not very wide, 0.1 to 10 microns is the norm, but they are very strong and protect the cell. They're also flexible to a point, just like a reed mat or a wicker chair. Thus, when water flows into a cell, pressure builds inside and the cellulose weave expands a bit. However, it can stretch just so far. This serves as a constraint to the buildup of internal water pressure. When the wall can't stretch, the inflow of water stops. This is turgidity. The full cell is turgid.

In addition to cellulose, other polysaccharides are produced and utilized in the cell wall. Synthesis of these requires a different set of enzymes. These help bind the cellulose together, forming cross beams, as it were. Pectin, the familiar Jell-O–like substance, is one such polysaccharide. The weaving together and gluing of the cellulose fibrils gives

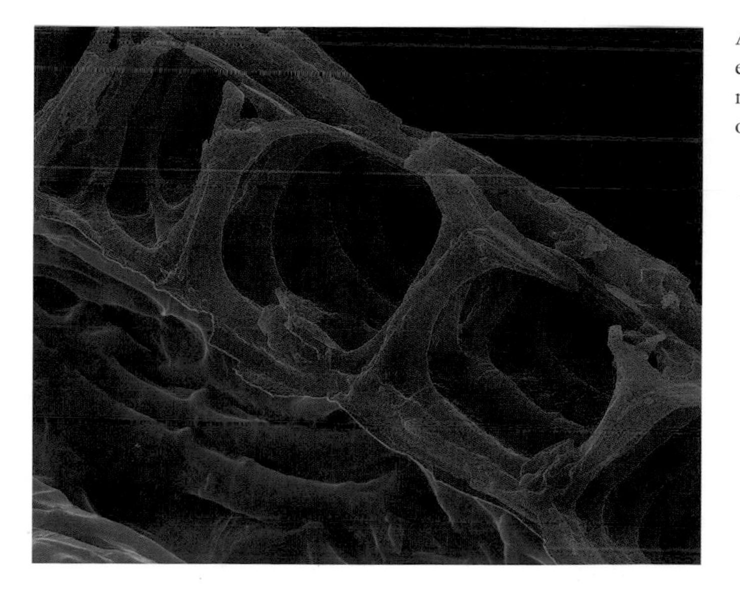

A scanning electron micrograph of cell walls

the cell wall great strength. Still, it is porous to small molecules, such as water.

The polysaccharide fibers that are exposed inside cell walls have negative electrical charges on them, and these attract positively charged particles, known as cations. Some important nutrients held in the soil are positively charged, and they are attracted to the negative charges on the polysaccharide fibers in the cell wall, which adsorb them (that is, hold them on the surface). These nutrients then become available to the plant.

Herbaceous plants just have a thin primary cell wall, whereas woody plants also possess a thicker secondary cell wall that provides additional support. This is a lignin-based layer. Lignin is another long-chained carbon molecule, and it is the major component of wood. A secondary wall sometimes sits up against the primary cell wall and usually contains suberin, a waxy, waterproof substance (again synthesized inside the cell). Cork is essentially suberin.

In both herbaceous and woody plants, there is a special set of cells located in the interior of the roots that separates the epidermis and the cortex from the area that protects the root's vascular tissue. This is known as the endodermis. Cells in the endodermis have walls with a strip of suberin and other waxes, which forms a barrier that prevents water from passing any further through that cell wall. A single layer of these cells encircles the center of the root, and the clogged portions of the cell walls line up to form what is known as the Casparian strip. It prevents water from flowing any further through the root and into the plant. To get there, water has to take a different pathway, one that will regulate what the water can bring into the cells along with it.

Water and gas molecules are small enough to travel around inside the porous cell wall and to flow through it into the adjoining cell walls (until the Casparian strip is encountered). This passageway into a plant through cell walls is called the apoplastic pathway, and it plays a key role in plant nutrition. Water in a cell wall or anywhere along the apoplastic pathway can move through the wall into plants without ever crossing over the cell membrane and entering a plant cell.

In sum, the cell wall is a porous, structural support for the plant cell. It is about 0.2 nanometer thick and sits up against the outer cell membrane. Because it is porous to water and gases, these small molecules can travel into plants by staying within the connected cell walls and not further entering the cells. Once water (and the molecules dissolved in it) encounters the waterproof Casparian strip in roots, however, it either meets the screening criteria of the next layer of the cell, the plasmalemma, and enters the interior of a cell or it travels out of the plant the same way it came in, via the apoplastic pathway.

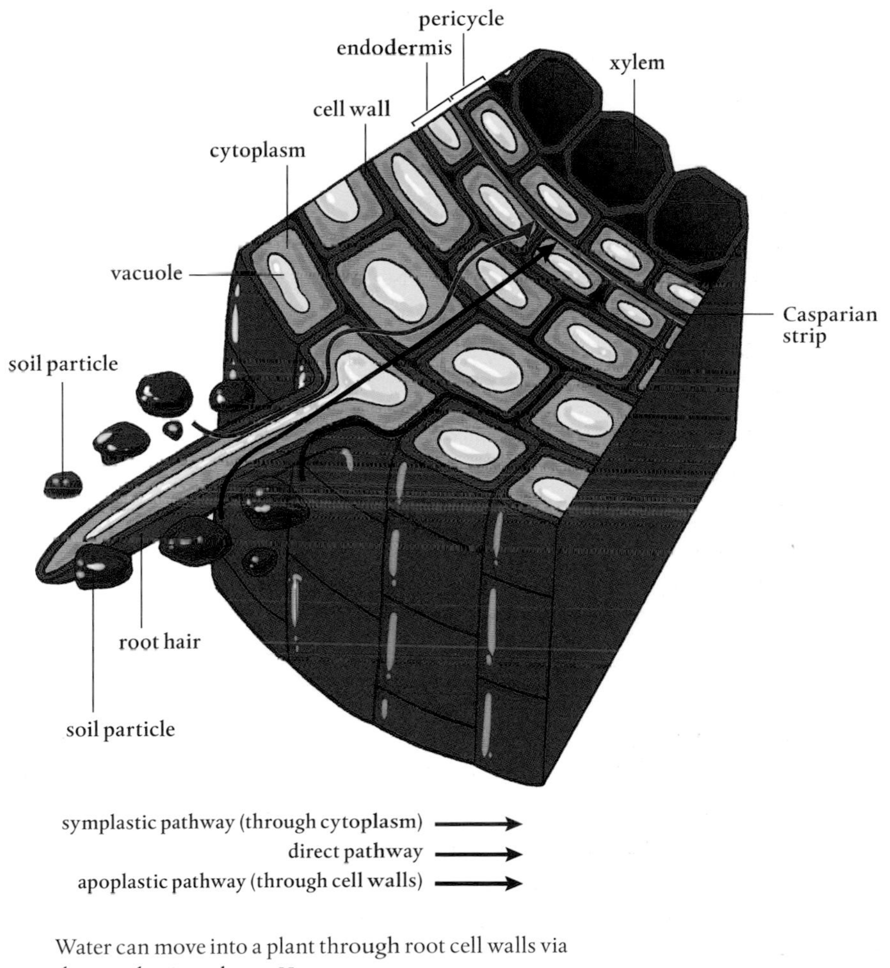

Water can move into a plant through root cell walls via the apoplastic pathway. However, once water encounters the Casparian strip, it must enter the cell across the cell membrane or move out of the root via the Casparian strip.

WATER PATHWAYS

Water moves through plants along three different pathways. The apoplastic pathway is located in the cell walls, which are porous and connected. Water can travel from wall to wall, but when it encounters the waterproof Casparian strip, it is prevented from crossing into the central area inside the root known as the stele. Water then either must leave the root via the cell wall system or move into a cellular pathway inside the plant.

The symplastic pathway is through the interior parts of cells. Water moves from the cell wall through the adjacent cell membrane and into the cytoplasm of a plant cell. Each cell of a plant is connected to its neighbors by tiny tunnels called plasmodesmata. The symplastic pathway is entirely inside plant cells, with molecules able to move from one cell to the next and then the next, throughout the plant. Cell organelles in the cytoplasm are not part of the symplastic pathway. They get in the way, and molecules must move around them if they are to stay on the symplastic path.

Finally, there is the intercellular or direct pathway. The chemical properties of water molecules and their small size allow them to move directly through cell walls and cell and organelle membranes, regardless of whether the structure acts as a barrier to other molecules. Thus, water can take the short route to the plant's vascular system, which is where all three pathways are headed.

Water is essential for the movement of nutrients through the plant. It is important to note that water molecules travel along all three pathways simultaneously, which helps to explain the tremendous volume of water that travels through a plant.

THE PLASMALEMMA

The plasmalemma is the outer cell membrane that sits against the inside of the cell wall. A permeable membrane allows things to pass through. Because only certain things can pass through the plasmalemma, however, it is a semi-permeable membrane, screening what can enter and exit a cell.

The plasmalemma is a barrier holding in the liquid cytoplasm and

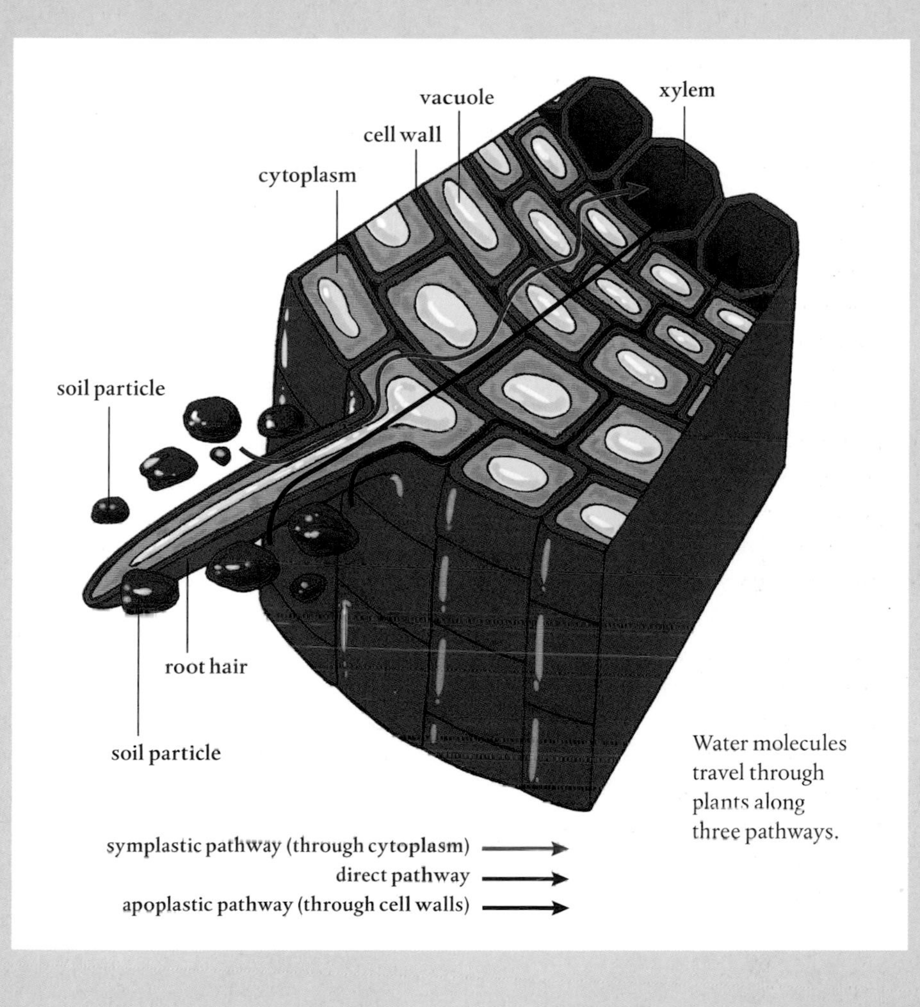

Water molecules travel through plants along three pathways.

organelles that make up a cell. The plasmalemma holds all of this inside, while keeping some other things from entering the cell. This is not just about getting nutrients in to the plant. The flow through the plasmalemma is bidirectional. How else would plant-produced exudates move out into the soil?

Plasma is matter that is not quite solid or liquid or gaseous, but rather in a phase that is something of all three. Many of you may have a plasma television set. Try moving your fingers around the screen when the set is on to see how plasma operates. Plasma fills fluorescent lamps and neon signs when electricity flows through them. Lightning is plasma, as are flames and the aurora borealis displays (hey, I am an Alaskan). The point is that the plasmalemma is a not a solid, rubber-like barrier. Rather, it is a semi-fluid barrier, much like a thick oil or half-jelled gelatin. Its constituent molecules are moving around and shifting places. It is essentially made up of two sheets of phospholipids, phosphorus-based molecules, punctuated with proteins and carbohydrates that have special functions.

Phospholipids are two-part molecules. The head is a negatively charged phosphate ion that is water-soluble. This means the heads dissolve in water. On the other end, the lipid tail is a carbon-based chain that lacks an electrical charge. It is not soluble in water. In fact, this end of the phospholipid is hydrophobic: it lumps together with other hydrophobic molecules rather than dissolving in water. This is exactly what happens to the fat in chicken stock or oil poured into water. It lumps together, rather than dissolving.

The plasmalemma consists of a double sheet of phospholipid molecules arranged so the water-loving heads point outward to the aqueous

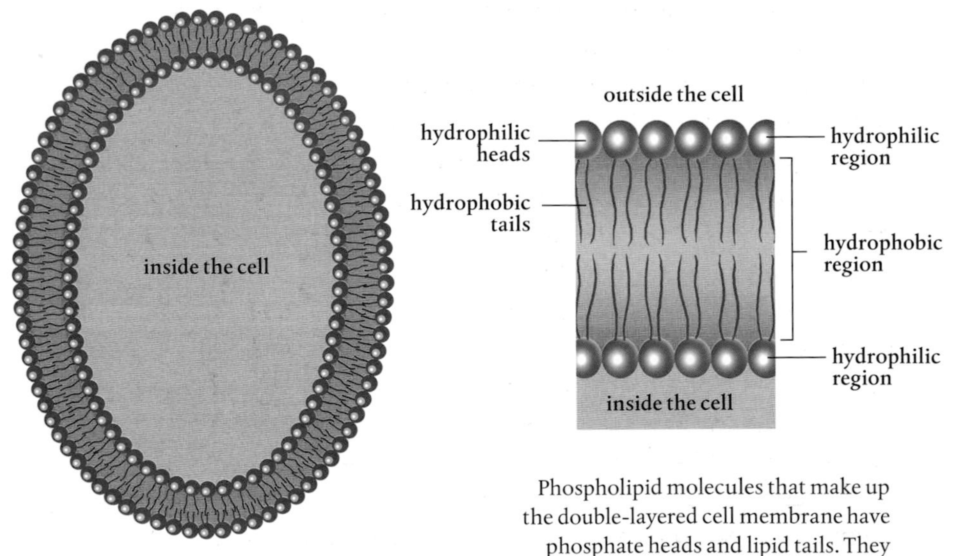

phospholipid cell membrane

Phospholipid molecules that make up the double-layered cell membrane have phosphate heads and lipid tails. They form a barrier that protects the cell.

extracellular environment and inward to the aqueous cytoplasm. The two sets of hydrophobic tails form an inner, middle layer that is not water-soluble. Imagine two pieces of bread each slathered with peanut butter, made into a sandwich. The peanut butter is the hydrophobic lipid tail layer and each piece of bread is a sheet of the water-soluble heads.

Certain molecules can easily cross through the plasmalemma. These are either small enough to fit between the phospholipid molecules or to dissolve in water and move through it in dissolved form. Molecules of oxygen, carbon dioxide, and nitrous oxide pass through the plasmalemma by diffusion, the natural movement of molecules to areas of lower concentration. Water molecules pass through the plasmalemma via osmosis, which is simply the diffusion of water.

How do other molecules get into or out of a plant cell if the plasmalemma is such an effective barrier? The answer lies in special proteins and carbohydrates embedded in and transversing the plasmalemma. These afford different kinds of molecules passage from the apoplastic pathway to the symplastic one, through the plasmalemma. They are key to how nutrients enter and wastes and exudates leave plant cells.

► **Plasmodesmata Tunnels** Plant cell walls are all connected via the apoplastic pathway. The area inside a cell membrane is connected to the inside of adjacent cells, forming the symplastic pathway, which is made possible by the presence of tunnels, plasmodesmata (or plasmodesma, when referring to just one). Because each and every cell of a plant is connected, if you were small enough you could start out in a root cell and eventually find yourself on the top of a leaf.

During cell division, bits of the plasmalemma get trapped between the two new cells and form the tunnels that connect the two. If you visualize a cell as one of those long balloons that can be twisted into an animal, you will note that bits of the balloon get twisted whenever another section is created. The balloon is like a mother cell's plasmalemma, and the twist becomes the connection between the two new cells. This happens every time a cell divides, such that all of the cells in a plant are connected internally. Therefore, it is possible for water and other molecules of the right size to travel from one cell to the next through these tunnels via the symplastic pathway. Because all the cells in a plant are internally connected by plasmodesmata, it is important that the plasmalemma screen out what should not be traveling into the symplastic pathway. Once a substance gets into one cell, there is a way to get to all other cells without being so regulated.

The analogy to the balloon twisting breaks down when you learn that an individual plant cell can have from 1000 to 100,000 plasmodesmata. These tunnels are 50 nanometers in diameter and up to 90 to 100 nanometers long. (Remember, a water molecule is about 1 nanometer in diameter.) Each has a smaller, tubular connection to the endoplasmic reticulum, a cellular organelle that serves as the pathway for transport of cellular materials inside a cell.

The size of a plasmodesma limits the size of the molecules that can pass through cells. At each end of these interconnecting tunnels (and sometimes inside them) are proteins that act like sphincters. These are responsible for regulating what molecules are allowed through. Again, there can be 100,000 of these little machines in a single cell membrane connecting it to other cells.

As for the differing numbers of plasmodesmata per cell, their numbers depend on the function of the neighboring cells. If there is a need for a lot of transport between cells, there can be as many as 15 plasmodesmata per square micron. (Many cells only have one plasmodesmata per square micron.) That's a cellular screen window, for sure.

► **Aquaporins** Water molecules are small enough that they can sneak through those lipid tails in the phospholipid membranes by osmosis, the movement of water molecules to where there is a lower concentration. When scientists measured the flow and amounts of water into and through cells, however, they realized that something other than simple water movement was at play. Aquaporins are special proteins embedded in the plasmalemma of plant cells that transport water molecules—and only water molecules—across the phospholipid membrane.

Aquaporins cover as much as 10 percent of the plasmalemma surface. Given the need for water in plants, this high proportion of surface area makes sense. As one might begin to suspect, plant cells involved in water absorption have more aquaporins than those that aren't. (Plants develop with amazing specificity that often passes by those who only observe the flowers.)

Aquaporins employ an electrical field to transport water from outside the cell through the plasmalemma, one water molecule at a time. Each water molecule has a positive end and a negative end, and the molecules can link up in chains. This polarity works with the electrical charge of the aquaporin protein. (Beam me up, Scotty!) Among other functions, aquaporins work with the plasmalemma and plasmodesmata to control the turgidity of the cell.

► **Transport Proteins Play a Crucial Role** Unlike plasmodesmata, which are actually part of the membrane, aquaporins are distinct and separate proteins embedded in the plant cell membrane. However, aquaporins are so strongly embedded in the membrane that it is extremely difficult to separate them from it. As such, they move with the plasmalemma. (It's plasma, remember? Imagine colors moving on a soap bubble.)

Aside from aquaporin, there are many kinds of transport proteins that are embedded in the plasmalemma. Each allows transit through the membrane of specific molecules. Together, these are known as integral membrane proteins (IMPs). What once appeared to be a smooth membrane has been discovered to resemble a rock climbing wall at the gym,

This atomic force micrograph shows the phospholipid bilayer plasma membrane and its proteins (spikes). The proteins either span the whole membrane (transmembrane proteins) or sit on its surface (peripheral proteins).

punctuated with hand and foot holds. These IMPs are the entryways for plant nutrients.

A tremendous amount of a cell's energy and the DNA blueprints it carries is devoted to making these transport proteins. Without each of them, a plant cannot take in the necessary building blocks for what it produces, and a cell makes everything it needs. That means there are a lot of different IMPs, as well as DNA to regulate their production and replacement. However, IMPs are key to any new resource intake, so the expenditure is well worth it—vital, in fact.

Integral membrane proteins can be grouped into three broad categories depending on how they transport specific molecules. The first group, channel proteins, serves as tunnels. Mechanisms at either end of the protein act as gates and regulate the flow of molecules by changing size or shape, thus giving rise to the name gated channels. Transit through channels does not require the addition of energy. It all depends on the natural propensity of molecules to move from areas of higher concentration to those of lower concentration, as well as the attraction between oppositely charged molecules. Because no external energy is required, this is known as passive transport. Don't confuse this with slow transport. Water and other specific ions pass through channel proteins, one molecule at a time, at the incredibly high speed of 10 million molecules per second.

The next group of IMPs consists of pump proteins. As the name suggests, these serve as miniature pumps. The most prevalent kind pumps positively charged hydrogen ions (H^+) out of the membrane, and these accumulate on the outer surface. Other molecules located there can use the energy of these hydrogen ions to move into the cell. There are sodium and potassium pumps. Because energy is used to run these pumps, their process is known as active transport.

Cotransporters or carrier proteins represent the third group of IMPs. These proteins bind to molecules, causing changes in the protein shape that move the bound molecules across the membrane. This is called facilitative transport because the transported molecule is helped or facilitated across the membrane. This is a form of passive transport because it does not require the input of energy.

Receptor proteins in the membrane transport signals across cell membranes. These membrane proteins cause chemical changes inside the cell as a response to extracellular stimuli. They act in concert with carbohydrate signaling mechanisms also found in the plasmalemma, forming infrastructure that connects all cells to one another and to the external environment. In many instances, receptor proteins create the chemical conditions inside or outside the cell that cause nutrients to pass into plants and they direct the nutrients' intracellular path.

Integral membrane proteins make up 50 to 75 percent of the surface of the plasmalemma. Remember, this membrane is already covered with plasmodesmata, as much as 10 percent of its area. These proteins provide part of the answer as to how nutrients get into plants and exudates and metabolites flow out. To move through an IMP, a molecule must be charged—that is, it must be an ion. Thus, plant nutrients are almost always in ionic form. The exception is boron (and organic molecules that apparently can be taken up by a small group of tundra plants).

Sometimes, there are materials that need to be moved through a cell membrane for which there isn't a specific transporting protein. Perhaps the molecule is too large. Endocytosis is the process whereby a membrane envelops something outside the cell, pinches off, and brings it into the cell. Conversely, exocytosis occurs when a membrane opens to the outside and dumps material out of the cell. Organelles inside plant cells and vesicles (something like a bubble) have membranes of the same material as the plasmalemma, enabling them to fuse to it. The material

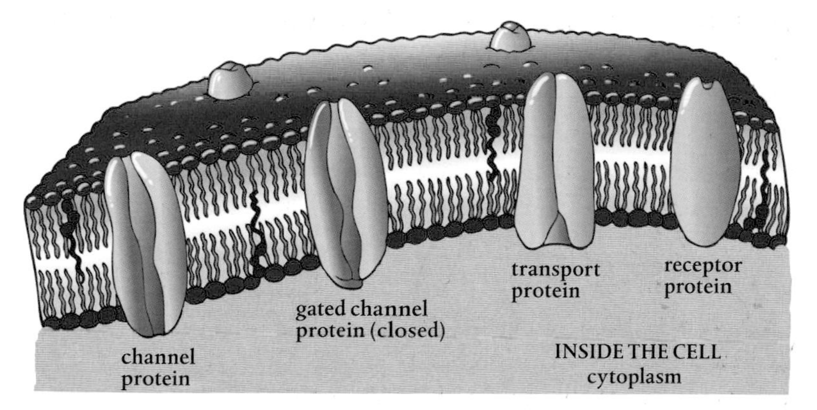

Cross section of the plasmalemma showing various types of integral membrane proteins

carried in a vesicle can be dumped outside of the cell by exocytosis without having to go through an IMP.

► **The Importance of the Plasmalemma** The plasmalemma is the regulator of what comes into and exits a plant cell. It enables the cell to keep a balance between the outside and inside, pumping water and ions in and out to ensure the proper conditions for life. The plasmalemma is the source of communication and signals between the cell and the outside world as well as with other cells. It helps to maintain and adjust cellular turgidity by regulating the amount of water held inside the cell. Finally, the plasmalemma is the site where the machines that build the cell wall operate.

CYTOPLASM

The cytoplasm is all of the stuff inside the cell membrane, except for the nucleus. The purpose of the membrane is to hold the cytoplasm in and keep unwanted things out. Cytosol, a clear, jelly-like fluid, is by far the major component of cytoplasm. It holds the cell's organelles, the sites of specific biochemical functions within the cell. The cytosol also holds vesicles that serve as tiny cellular warehouses. Some of these store building blocks such as carbon, oxygen, and proteins. Others act as cellular trash-cans containing wastes and sequestering toxins until they can be disposed of outside of the cell, mostly by exocytosis.

MITOCHONDRIA

Mitochondria are organelles that produce energy for the cell from the sugars made during photosynthesis. Because cells require so much energy, there are hundreds and even tens of thousands of mitochondria in each plant cell. The mitochondrion is the site of cellular respiration, the breaking down of sugar using oxygen. The energy formed in the process is stored in adenosine triphosphate (ATP) molecules, the universal currency for cellular energy.

Mitochondria are bound by their own double phospholipid membranes, complete with embedded transport proteins. However, the mitochondrion's inner membrane is folded over and over like ribbon candy, producing numerous attached compartments, or cristae. This

MITOCHONDRIA

Mitochondria are miniature generators, the powerhouses of the cell. They are the sites of respiration, the process that produces energy from the sugars made elsewhere. A typical plant cell can have tens of thousands of these cellular generators. The number appears to depend on the cell's function in the plant.

The cylindrical mitochondrion has a double-layered phospholipid membrane, complete with embedded transport proteins like the plasmalemma of the plant cell. The big difference is that the inner membrane is folded to increase the surface area and to create compartments, much like ribbon candy. These compartments are cristae, the sites where the necessary chemical reactions take place. The energy created by these biological generators is stored in the phosphate bonds of ATP molecules.

Mitochondria not only have a double phospholipid membrane typical of plant and animal cells, but they also have their own DNA and divide just like bacteria. In fact, it appears that these organelles once were free-living bacteria that entered into such successful symbiotic or parasitic relationships with plant cells that they were finally biologically assimilated into them.

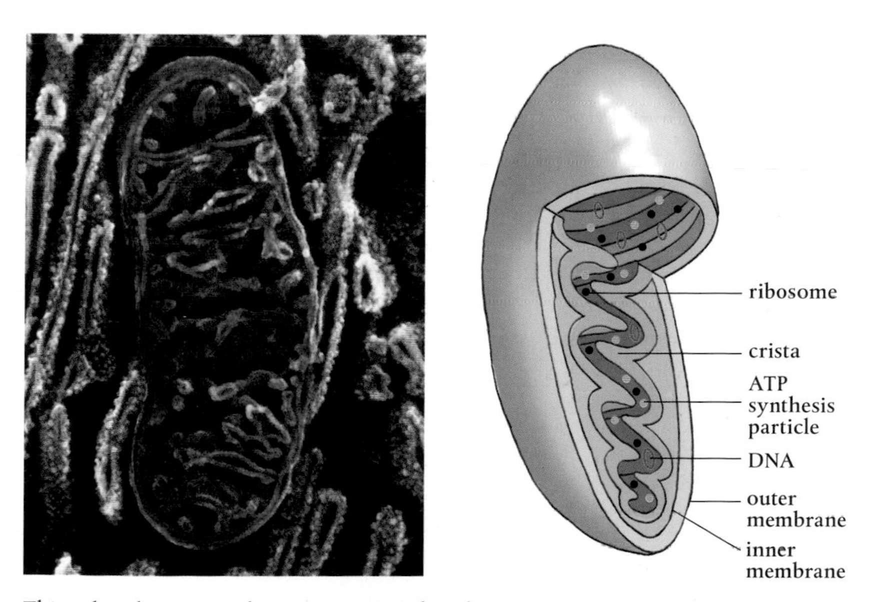

This colored scanning electron micrograph and accompanying diagram each show a single mitochondrion.

folding provides more surface area so more energy-producing chemical reactions can take place.

Inside the cristae, enzymes strip electrons from glucose molecules and use them to add phosphorus atoms to adenosine diphosphate (ADP) molecules, making ATP molecules. The energy in the new phosphorus bond is what makes ATP so valuable. Breaking this bond releases the energy. As you might expect, cells that require more energy have more mitochondria to produce more ATP molecules.

Producing energy is not the only role of mitochondria, however, They also synthesize the phospholipids that are so necessary for cellular membranes. Mitochondria also help to store calcium, which is used in intercellular signaling and directing traffic in the symplastic pathway.

To the studied eye, mitochondria look a lot like bacteria. They carry their own set of DNA, which is similar to that of bacteria. They divide like bacteria and have a double membrane, which is just like that of bacteria. Scientists now believe that mitochondria were once free-living bacteria that somehow got into the plant cell. There is an argument as to whether these bacteria entered into a symbiotic relationship with the plant cell after being ingested by endocytosis, or whether the remnant DNA that shows the ancestral bacteria had a flagellum (a moving tail that provides locomotion to certain bacteria) means the early relationship was more parasitic.

In either case, there is little argument that eons of assimilation have made mitochondria an integral part of plant cells. Without them, there simply wouldn't be enough energy to get things done and no plasma membranes to hold, protect, and regulate cellular content. Fortunately, you can find them all over the cellular factory, generating power for plant life (and generating phospholipids in the off hours). Mitochondria account for up to 20 percent of a cell's total volume. Life, no matter how tiny, runs on energy.

PLASTIDS

Plastids are miniature factories and storage facilities that are found throughout the plant cell. There are three groups of these organelles: chloroplasts, leucoplasts, and chromoplasts. Chloroplasts are the sites where photosynthesis takes place. They are filled with chlorophyll, a green pigment that absorbs energy from sunlight. The light excites electrons, and this energy is captured. Chlorophyll gives plants their green color. Leucoplasts are colorless (or white) and are found in parts of the plant not exposed to light, such as the roots and seeds. They are where starch grains, lipids, and proteins are formed and stored. Leucoplasts are also found in the white or colorless parts of variegated leaves, suggesting that they may be mutated chloroplasts. Amazingly, some leucoplasts can convert into chloroplasts when exposed to light. Chromoplasts contain the red, yellow, and orange pigments found in flowers, fruits, and some roots. These colors are useful in attracting pollinators and seed dispersers.

Chloroplasts are the most familiar of the plastids and are shaped like tiny lenses (2 to 8 microns long). An average plant cell contains 10 to 100 of them. The average leaf (if there is such a thing) would have a whopping 500,000 chloroplasts per square millimeter. Like mitochondria, chloroplasts have a double phospholipid membrane. The inside membrane is also folded to add a lot of surface area for the photosynthetic reactions to take place. The concept of a conveyor belt comes to mind.

Inside each chloroplast are thylakoids, membrane-bound structures that are flattened, hollow discs, stacked one on top of another. Attached to these discs are green, chlorophyll-filled antennae that absorb light. When light hits the chlorophyll molecules' electrons, they jump to an outer orbit (I will discuss electron orbits in the next chapter),

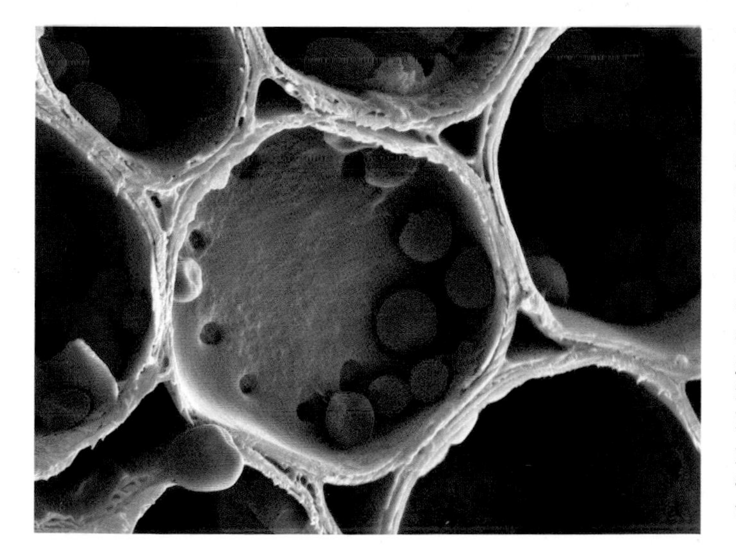

This scanning electron micrograph shows starch grains within amyloplasts in the cell wall compartments of potato tuber cells. Starch is synthesized from glucose, a sugar formed in the leaves during photosynthesis and transported to the tuber.

which increases their energy level. The souped-up electrons flow like electricity through the thylakoid and accumulate in one membrane compartment.

When the concentration of these electrons inside the thylakoid compartment becomes greater than outside, the laws of thermodynamics take over and they fight to move out and away from each other. They can do this only by passing though a special channel protein in the membrane, the enzyme ATPase. The movement of electrons through the channel causes phosphates to be added to ADP molecules, creating ATP, which is the currency of energy in biological systems. Removing the phosphate yields ADP again, plus the release of energy.

The electrons then get a second dose of light and are picked up by nicotinamide adenine dinucleotide phosphate (mercifully abbreviated as NADP). The ATP created in the first go around is used to form sugar molecules. Each chloroplast can create thousands of these every second, provided there is light to excite electrons in the chloroplast. Remember, there are half a million chloroplasts in a square millimeter of leaf. That is quite a sugar factory.

There is a close similarity between chloroplasts and cyanobacteria, chlorophyll-filled photosynthesizing bacteria. Both have a double membrane and DNA and both divide. Again, this similarity suggests that some ancient, free-living ancestors were ingested into plant cells or entered as parasites. It is evident that the cell incorporated parts of the cyanobacterial DNA into its own, and chloroplasts are now an integral part of plant cells.

RIBOSOMES

Ribosomes are bead-shaped organelles that serve as the sites of protein synthesis, which requires much of the energy produced by a cell. They are the smallest organelles in a plant cell (25 to 30 nanometers). Ribosomes can be attached to the membranes of the rough endoplasmic reticulum (another organelle described later in this chapter) or floating free in the cytosol. Although some cells have thousands of ribosomes, most usually only contain about a thousand. Perhaps it is due to their size that ribosomes were not discovered until 1953.

Ribosomes are constructed in the nucleus and transported out into the cytoplasm. Ribosomes are interchangeable, and their numbers

CHLOROPLASTS

There are typically forty to fifty chloroplasts in most plant cells. Here, each chloroplast is shown as if cut lengthwise. Each is surrounded by an external double membrane, which provides protection for the operations inside. The area surrounded by the inner layer is known as the stroma. This is where light reactions occur during photosynthesis. Located here are the grana (yellow), stacks of flattened thylakoids. Only grana contain the chlorophyll pigments that give plants their green coloration and convert sunlight to energy. Light energy absorbed by the chlorophyll in the chloroplast is used to convert carbon dioxide and water into sugars and starches. The large amounts of starch produced during photosynthesis appear as dark circles within each chloroplast.

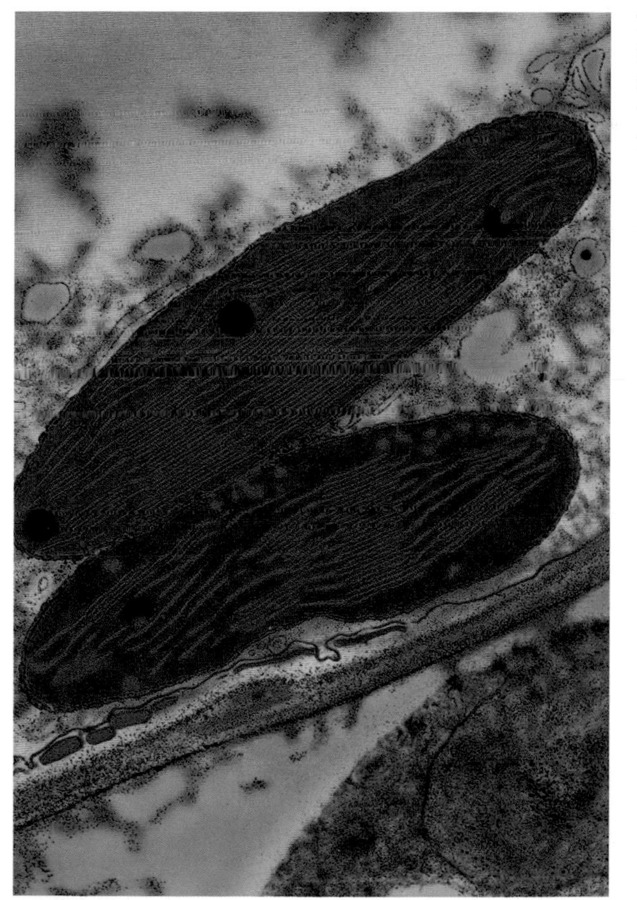

This colored transmission electron micrograph shows two chloroplasts in the leaf of a pea plant.

fluctuate depending on the need for proteins. The free-floating ribosomes make proteins for use in the cell, whereas those on the endoplasmic reticulum synthesize proteins that are exported out of the cell for use elsewhere in the plant. There are no lipids in ribosomes, no membranes. Ribosomes also carry a negative charge.

Imagine the ribosome as a very busy knitting machine, one that you pull balls of wool through and out comes a woven strand. Into the ribosome go strings of amino acids, the building blocks of proteins, and out they come connected in just the right order to make a specific protein. A ribosome can make various kinds of proteins depending on the chemical signals it receives. The particular combination and order of the amino acids (there are twenty) determines the characteristics of the protein.

Deoxyribonucleic acid (DNA) and ribonucleic acids (RNAs) are the molecules involved in replicating the genetic code used to build everything in a cell. The instructions for the particular protein are delivered by messenger RNA (mRNA), and the sequence is copied in the ribosome. The amino acids that are made into proteins are brought to the ribosome by transfer RNA (tRNA). A well-fed plant will have many

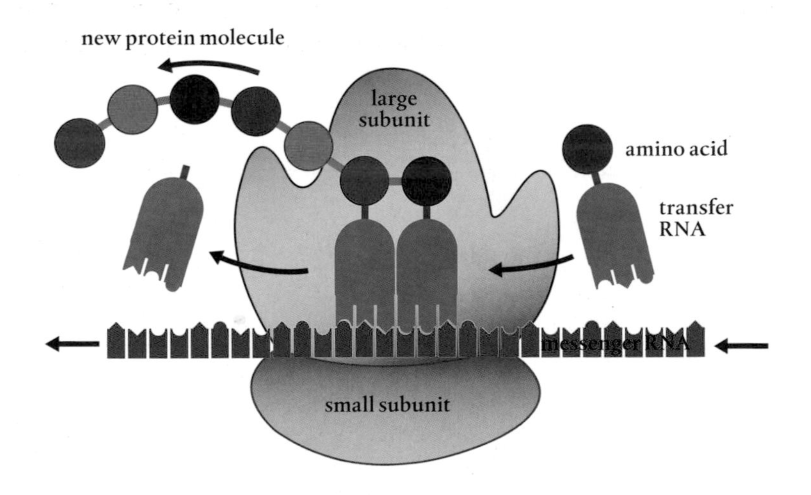

Ribosomes are the sites where messenger RNA sequences (copied from DNA in the nucleus) are read and proteins are synthesized from combinations of the twenty amino acids. The amino acids are brought to the ribosome by transfer RNA.

times more tRNA than mRNA, so there is never a need to wait for supplies to construct a protein.

Don't be fooled by the tiny size. Each ribosome is a real assembly line, precise in its actions, with all the equipment necessary to connect amino acids in the proper order to synthesize proteins. They are small, but the synthesis of proteins uses much of the energy produced by the cell. In any case, some of these proteins are used inside the cell, and others are exported to different parts of the plant. Once a protein chain is created in the ribosome, it is released and transported to the Golgi apparatus, another organelle inside the plant cell, for final processing before being taken to the cell membrane or used inside the cell.

THE ENDOPLASMIC RETICULUM

The endoplasmic reticulum is a long membrane folded over itself again and again to create the lumen, a maze of cavities. Many ribosomes are attached to a section of these folds, making it look bumpy, hence it is called the rough endoplasmic reticulum.

The number of ribosomes thin out along the smooth endoplasmic reticulum, and it is from here that the proteins produced in the ribosomes are shipped to other parts of the cell. Nucleotides are added to the ends of the chains to serve as the address for delivery. The smooth endoplasmic reticulum is also the site where proteins have sugars added to

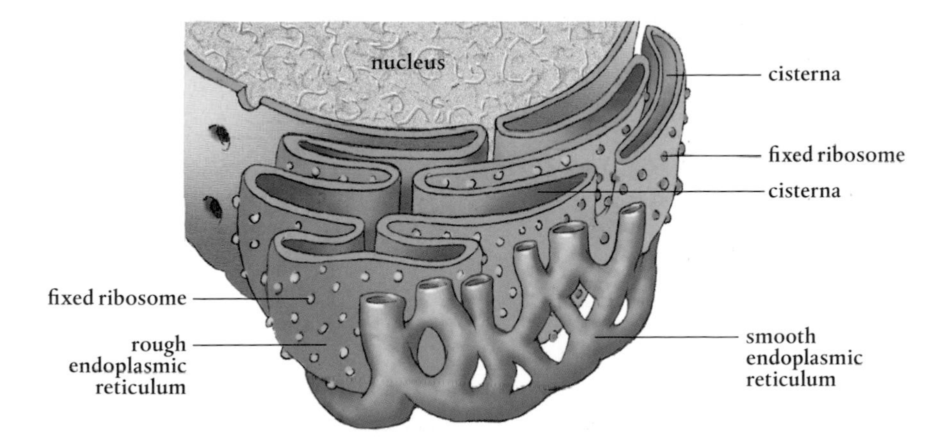

The endoplasmic reticulum extends from the nucleus to the cell membrane.

them to become glycoproteins needed for certain cellular functions and reactions. In addition, lipids are made in the smooth endoplasmic reticulum. Another important function of this organelle is changing toxins to more benign substances so they can then be safely transported out of the cell. Vesicles form by endocytosis around waste toxins here, and vesicles also transport proteins from the endoplasmic reticulum for further processing.

The endoplasmic reticulum extends from the nucleus throughout the cell to the plasmalemma. The extensive network of the endoplasmic reticulum allows molecules to travel an easier path than having to slog through the jelly-like cytosol and dodging all manner of intercellular molecular traffic of synthesized and synthesizing molecules.

THE GOLGI APPARATUS

The endoplasmic reticulum is connected to the Golgi apparatus (or Golgi complex), membrane-enclosed organelles named after an Italian histologist, Camillo Golgi. (I can't think of anyone else whose name graces a cell organelle.) These are the packaging and shipping centers in a cell. Each Golgi apparatus is surrounded by tiny tubules that start at the endoplasmic reticulum and radiate throughout the cell. These and less visible axial fibers are used as tracks along which

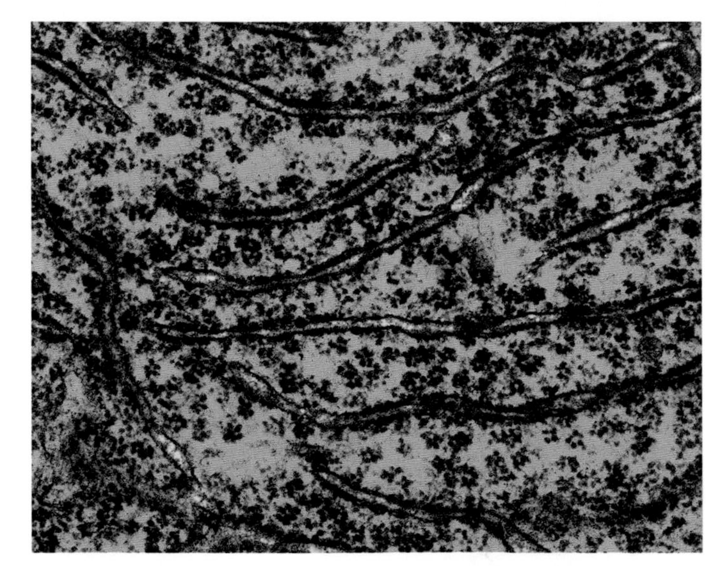

Golgi apparatus in a plant cell. Stacks of cisternae and vesicles are present.

loaded vesicles move. The vesicle fuses with the membrane of the Golgi apparatus and deposits its contents into the assembly line, where they are modified.

The Golgi apparatus is really a number of Golgi bodies, stacks of membrane-enclosed discs. These contain enzymes that modify the materials sent to the complex. Once finished, the Golgi apparatus ensures these materials reach their ultimate destination, which can be inside or outside the cell. What kind of materials? Proteins are finished in the Golgi apparatus, and the polysaccharides (long sugar chains) needed for making cellulose and used in root exudates are assembled there as well.

There are five to eight Golgi apparatuses in a plant cell. This is not a lot when you realize that almost all the molecules in a cell pass through one of them. Each Golgi body within a complex contains separate sets of enzymes. Inside the Golgi bodies are long passages where molecules can be finished, sorted, chemically tagged with a destination code, and loaded into vesicles for final transport. The cisternae in the sacs contain lots of tiny transport vesicles to which specific proteins and polysaccharides bind for the transport out. Each vesicle's phospholipid membrane merges with the cell membrane and the contents are emptied. Those vesicles containing waste material fuse with the vacuole.

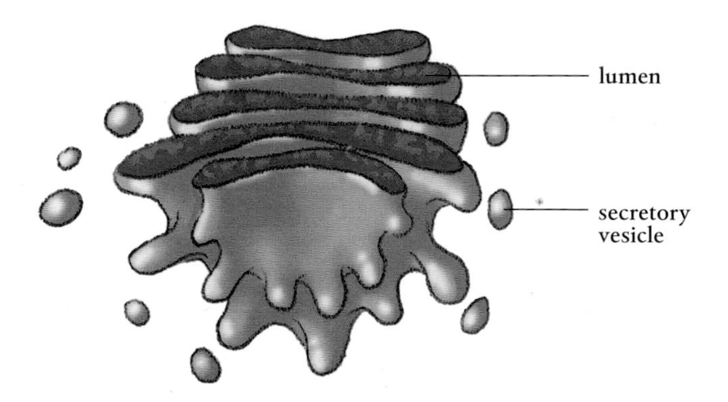

The Golgi apparatus is where the final packaging is done, and then molecules are shipped out in vesicles.

VACUOLE AND TONOPLAST

The vacuole is a single-membrane organelle filled with cell sap, a watery mixture of molecular compounds. This sap helps maintain pressure (turgor) in the cell. Cell sap mostly contains water, as well as sugars, enzymes, proteins, organic acids, and the pigment molecules responsible for the red, blue, or purple of flowers and the red of leaves. It also contains cellular wastes. It is slightly acidic, meaning it has a pH below 7.

Young plant cells may have several small vacuoles. These individual vesicles have the same kind of plasma membrane, so they can merge, just like smaller soap bubbles merge to become larger ones. Mature cells have only one large vacuole. The vacuole's membrane is the tonoplast. Like the plasmalemma, the tonoplast is embedded with specialized transport proteins that allow materials to move across the membrane without having to deal with the barriers created by the phospholipid molecules. Transport of materials across the tonoplast most often requires the input of energy.

Each plant cell usually has one main vacuole, which can make up as much as 90 percent of the cell's volume. The vacuole fills and increases the turgor, or pressure, within the cell. A full vacuole is one reason plant stems can support heavy fruits and store water. When the vacuole shrinks enough due to a shortage of water, the plasmalemma pulls away from the cell wall. Collectively, this causes plasmolysis or wilting.

The tonoplast helps keep the pH in the cytosol at the proper level.

Epidermis cells after plasmolysis. The vacuoles (stained pink) have shrunk.

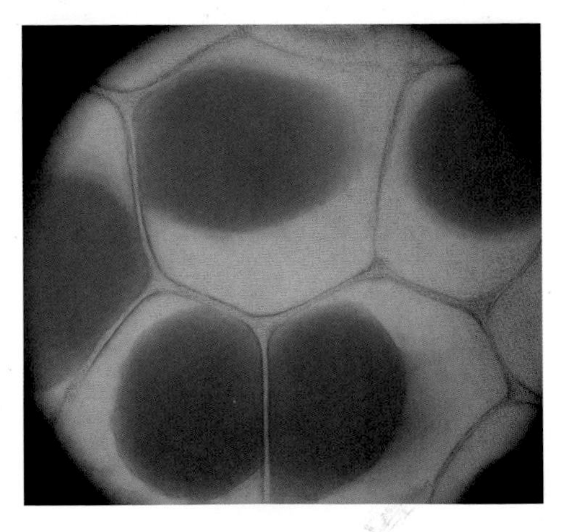

The pH is a measure of the concentration of hydrogen ions (H^+) in a solution: the more of these, the lower the pH and the more acidic the solution. Some of the transport proteins in the tonoplast membrane can pump hydrogen ions into the vacuole, thus increasing their concentration and lowering the pH. The concentration of hydrogen ions is carefully regulated to keep the pH in the cytosol around 7, which is neutral.

As for the wastes collected and deposited into the vacuole, in some instances part of the tonoplast merges with the plasmalemma to export wastes from the cell, or wastes are moved to the lysosomes and peroxisomes. Cell death is usually a result of the tonoplast no longer working. When it fails, wastes are no longer contained, turgor is affected, and the pH of the cytosol is altered.

LYSOSOMES AND PEROXISOMES

The vacuole membrane can also merge with lysosomes and peroxisomes. Lysosomes are the cell's recycling centers. They are formed when part of the Golgi apparatus membrane buds off. Lysosomes contain enzymes that digest large molecules (mostly proteins) into their components, which are then recycled in the cell. The number of these digesting vessels varies depending on the cell function, but can range up to 100 or so per cell.

The enzymes in lysosomes require an acidic pH around 5 to do their job. If the pH is any higher, they won't work. The pH of the cytosol needs to be close to neutral so the other organelles and their products are not harmed. This means a lot of hydrogen ions must be pumped into lysosomes to keep the pH low enough for their enzymes to function. Should some of the enzymes leak out of the lysosome, the higher pH they encounter prevents them from functioning and digesting organelle membranes and materials needed for synthesis. At the end of a cell's life, a signal is sent to lysosomes to release their enzymes to digest the dead cell. Apparently, without cellular function, the environment becomes acidic enough so the enzymes can work.

Lysosomes only have a single-layer membrane, which means they can more easily merge with the membranes surrounding vesicles and vacuoles, enabling them to receive waste that is then taken away and broken down. This waste stream includes bacteria, nutrient molecules, proteins, lipids, polysaccharides, and nucleic acids. It can even consist of organelles that are no longer functioning, such as a ribosome or spent mitochondrion. All of the material is recycled.

Peroxisomes also serve as digestion vessels in the cytosol. These are generated from the proteins and lipids made in the endoplasmic reticulum. Unlike lysosomes, the enzymes in peroxisomes digest lipids and fats. In seeds, for example, they supply the enzymes that start the conversion of stored fatty acids into sugars. Peroxisomes also help in the cellular assimilation of nitrogen and the metabolism of hormones.

THE NUCLEUS

The nucleus is the only organelle that is not considered part of the cytoplasm. It is here that a plant cell's hereditary material, its genome, is stored along with instructions on how to synthesize proteins. The nucleus is the coordination center for cellular activity, and it accounts for about 10 percent of the space in a cell.

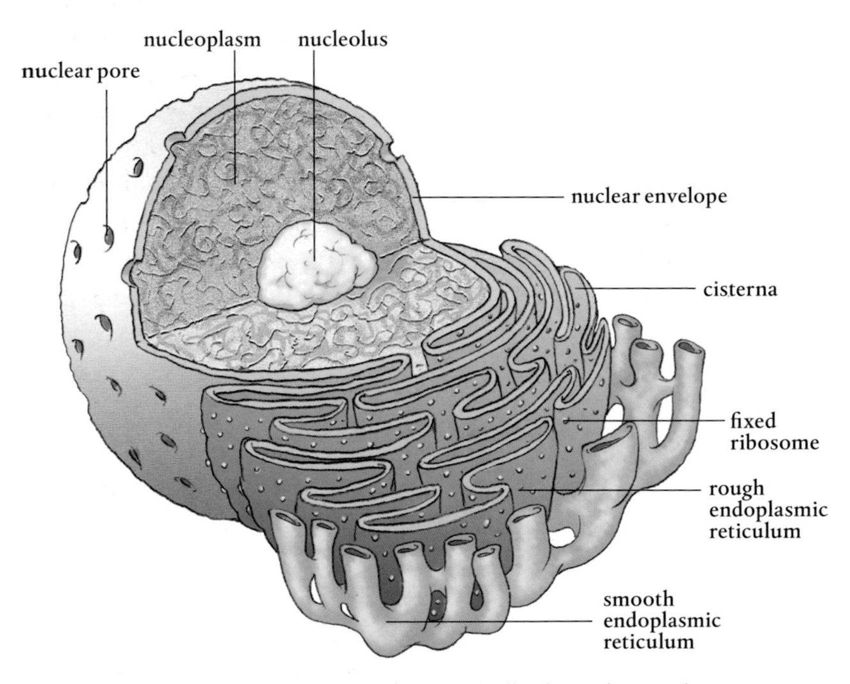

Arguably the most important organelle, the nucleus is where DNA is stored and protected. Note the connection to the endoplasmic reticulum and the nucleolus, where ribosomes and RNA are assembled.

The nucleus is surrounded by two phospholipid membranes that are separated by a space known as the nuclear envelope, which surrounds the entire nucleus. The outer nuclear membrane is continuous with the membrane of the rough endoplasmic reticulum and is also studded with ribosomes. These membranes keep the cellular cytoplasm separated from the stuff inside the nucleus, the nucleoplasm or nuclear sap that includes DNA and RNA.

There are holes in the nuclear envelope that cover up to 15 percent of its surface. These nuclear pores range in diameter from 10 to 100 nanometers. They allow the nucleus to communicate with the cytosol while screening and regulating the molecules trying to transit the nuclear membrane in either direction. Small building-block molecules have no problem moving through. However, large proteins and RNA have to be selected to move through these pores. Obviously, there is an important need for control of what gets in and what is let out of the nucleus. After all, it holds the DNA of the cell and can't be contaminated.

The nuclear lamina, the inner membrane of the nuclear envelope, contains a mesh network of fibrous proteins that anchor the nucleus in the cell and give it some rigidity. Inside the nucleus are the building blocks, the cell's precious DNA and RNA, and the nucleus's own organelle, the nucleolus. This is where the units that make up ribosomes are produced, along with the RNA molecules used for duplication of the DNA and transport of amino acids during protein synthesis. Both are transported out of the nucleus to the endoplasmic reticulum and moved on to the Golgi apparatus. Most plant cell nuclei have more than one nucleoli.

MICROTUBULES AND ACTIN FILAMENTS

Most plant cell organelles are connected to or in extremely close proximity to a network of transport tubules known as microtubules and actin filaments. These are 2000 times thinner than a human hair. Together actin filaments and microtubules provide structural support for the cell and serve as its transportation infrastructure and communication network. Molecules move along the microtubules and filaments as if they were railroad tracks. Signals can be sent through them and along them as well, so they serve as the intracellular Internet or telegraph system.

It may seem duplicative to have two kinds of organelles perform essentially the same functions. However, when it comes to being able to send signals and deliver the building blocks, finished products, and organelles necessary for survival, redundancy makes a lot of evolutionary sense. In addition, the microtubules and actin filaments work together, so there is synergism between the two systems.

THE BIG PICTURE

It does take imagination to view plant cells in your mind. How do you get your mind around the fact that there are about 10,000 different kinds of proteins moving around in each microscopic cell, with as many as 10 billion individual ones in total? Then there are the precursors to proteins as well as almost countless numbers of other molecules made by the cell. The number of molecules inside a plant cell, frankly, is incomprehensible, and moving through them must be like swimming in a stew.

Most of us will never have access to the kind of equipment it takes to see the molecules in a plant cell in any greater detail than we did in

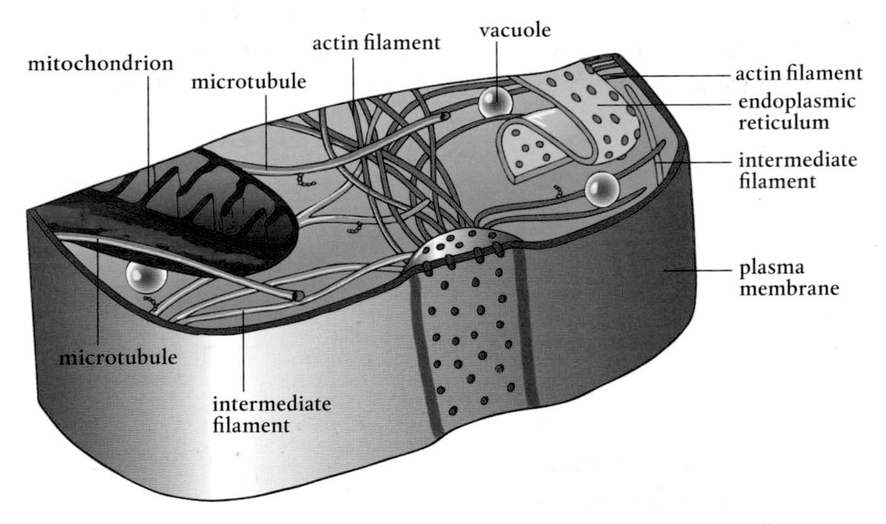

There are thousands of tiny filaments in a cell. Actin filaments are the smallest, followed by intermediate filaments, and larger filaments known as microtubules. These elements form a transportation system, a set of tracks, along which things can be moved around the cell.

biology class. The sizes are so small that the measures are not ones to which the gardener has a reference point. So, we're left with imagining what things are like in a cell.

Let's try to put things in perspective by reducing ourselves down to the size where we can really relate to a cell. If a cell were 300 million times larger, the nucleus would look like a sponge ball about 50 feet across. Depending on where we are located, at this diminutive size an entire cell looks like a football stadium.

Unfortunately, due to all of the snowflake-sized molecules floating around, laced with hail-sized proteins, it is difficult to see anything clearly. Actin filaments are the diameter of a salt shaker, the intermediate filaments are halfway between that and the larger microtubules, which are the diameter of a beer mug. They hang about like cobwebs, and they are everywhere.

Speaking of cobwebs, tarantula-sized ribosomes appear all over the place, floating around in the jelly-like cytosol. And you can see them hanging on the rough endoplasmic reticulum, hundreds and hundreds of them. The endoplasmic reticulum is huge, extending from the nucleus to the plasmalemma, its sacs about as wide as an adult hand. While it extends the length of the cell, it only covers about a third of the cell's width. It is almost like a slide, protected from the jelly-like substance that fills the cytoplasm. It is much easier to travel the endoplasmic reticulum than in the cytosol.

Mitochondria are now the size of tractor-trailers. You may see some being pulled along the microtubule network. Lysosomes, grinding away at their contents, are spheres that range from the size of softballs and soccer balls to huge medicine balls with diameters up to three times the size of a human. That Golgi apparatus you almost stepped on is twice the size of your offending foot and about as wide. You wonder where the other four or five are, so you won't step on them, too.

You check out the apoplastic pathway by wandering around the cell wall system. It is surprisingly like thin cardboard or sliced luffa sponge. You can see the cellulose fibers, like wires; they are extremely strong.

You find the plasmalemma. This membrane is thicker than the cellulose fibers, but not by much; it has the thickness of this book. And it is constantly flowing, like oil in water, only it has all of these shoe-sized proteins sticking through it along with rope-like chains, which must be some sort of signaling molecule. Watch out. Don't get sucked into an aquaporin or shocked by some of the peanut-sized charged hydrogen ions that are affixed to the surface of the cell wall.

Glowing ATP molecules are breaking apart all around, bursting into sparks like fireworks, and then the glowing stops as their energy is spent. Everywhere there are enzymes that are constantly changing shapes, attaching to and detaching from other molecules. The activity is overwhelming, and the noise is fantastic. Things are forming everywhere in incredible numbers and then being transported. Motion is constant.

Somehow, all of this activity moves forward. Energy is supplied, and if the nutrients are available to be converted into construction material, a tremendous amount of synthesis goes on inside the cell—building, repair, changing construction as a result of a signal sent through a carbohydrate molecule running through the cell membrane. Multiply this activity by 15 trillion and you get a birch tree. It's amazing.

To be able to really see what goes on in a cell would be mind boggling. You probably wouldn't be able to take too much of it and would want to get back quickly to read the next chapter in the story of how plants eat. We will continue this story by moving on to some chemistry and a bit of botany. All will be connected, and things will become clearer.

TWO

Some Basic Chemistry

WHAT'S IN THIS CHAPTER

- **Each nutrient** is an element, a substance composed of identical atoms.
- **An atom** has positively charged protons in its nucleus, which is surrounded by a cloud of negatively charged electrons.
- **When atoms** of the same or different elements share electrons, a chemical bond is formed.
- **Nutrients form** three types of bonds: covalent, ionic, and hydrogen bonds. These affect the qualities and availability of the various nutrients.
- **Plants take in** the nutrients in order to build the four types of molecules that are necessary for life: carbohydrates, proteins, lipids, and nucleic acids.
- **Adenosine triphosphate** (ATP) serves as the energy currency in cells.
- **Enzymes are proteins** that dramatically speed up the millions of chemical reactions that occur in plant cells.

- **Osmosis and diffusion** are two processes that allow water and nutrients to move across membranes without the input of energy.
- **The active transport** of nutrients across a cell membrane requires energy.

CHEMISTRY DEALS with atoms and molecules and how they react with other atoms and molecules. That's all chemistry is. These behave very predictably. Yes, it is chemistry, and many people have a fear of it. Just take a deep breath or two and dive in. There won't be any tests or grades. You don't have to memorize. "You will learn," as my mother used to incorrectly say, "by osmosis." Pretty soon you will have enough familiarity with the necessary elements. Don't let the chemist's vocabulary get in the way of learning.

ATOMS

All substances are made up of atoms. An atom is the smallest unit of an element that retains the properties of that element. Atoms consist of a central nucleus that has a certain number of protons, which are positively charged particles, and neutrons, which have no charge, depending on which of the 118 elements the atom belongs to. The atom's nucleus is surrounded by lighter particles known as electrons. These carry a negative electrical charge.

Atoms are often depicted as planets with circling moons, with the planet representing the nucleus comprising protons and neutrons and the moons representing the circling electrons. The orbiting planet view of an atom is an oversimplification of atomic structure (and I am all about simplification). The electrons actually form a swirling cloud or shell around the entire nucleus, but for our purposes the planetary model suffices.

There is an electromagnetic force between the positively charged protons in the nucleus and the negatively charged electrons surrounding it that keeps the electrons in orbit. An atom can have several orbits of electrons, with each consecutive orbit having a higher energy level.

By definition, atoms are small, a whole new level of smallness, in fact. If an orange were the size of the Earth, each atom in it would be the size of an orange. Or imagine that a pea represents the size of an atom. The cell we just studied would be the size of an ocean-going cargo vessel. Just remember, as small as it is, an atom still has three dimensions.

Each element has a unique number of protons, its atomic number. For example, hydrogen has one proton in its nucleus, and nitrogen has seven. But for the most part, and certainly for our purposes, it is the number of electrons in the outer orbit of an atom that determines its chemical behavior.

Electrons like to pair up whenever possible. Most of the chemical behavior that gardeners care about is a result of pairing up all of the electrons in the outer shell of an atom. This is the driving force of chemistry and, indeed, life on an atomic level. (Indeed, many would argue, it's the driving force of life on any level.) The outer shell of most atoms can hold four pairs of electrons, or eight electrons in total. This is called an octet, and Lewis diagrams are used to show the electron configuration around particular atoms and combinations of atoms in pairs or as loners.

The outer orbit of electrons of an atom is known as the valence orbit. Sometimes there are less than all eight electrons in the shell, but there are never more than eight. Because electrons tend to form pairs, an atom with fewer than four pairs in its valence orbit is highly reactive. In a sense, an atom seeks out the electrons of other atoms in order to pair off the electrons in its outer orbit.

Chemists use shorthand to represent elements and the combinations they make. Thus, water is H_2O. This means that each water molecule consists of two hydrogen (H) atoms and one oxygen (O) atom. It is the completion of the pairing of electrons in the valence orbits of the hydrogen and oxygen atoms that results in chemical bonding of the three and the creation of water.

MOLECULES AND COMPOUNDS

Because valence electrons like to share their orbits, single atoms rarely occur. Instead, when vacancies appear, other atoms come to complete the pair. This causes the atoms to bind together to form molecules. A molecule is the smallest unit of a compound substance that retains the properties of that substance. Just as an atom can't be further divided, if you tried to divide a molecule of a compound, you would end up with its constituent atoms. If the atoms are the same, the molecule is an elemental one (such as nitrogen, N_2, which is composed of two nitrogen atoms bonded together). If they are different, the combination results in formation of a compound (such as water, H_2O). In every case, however, it is the electrons in the outer shell that do the bonding and create the chemistry.

Electrons can behave in different ways, which results in different kinds of bonds, each with special characteristics that give the molecules different properties. There are just three kinds of bonds of importance: covalent, ionic, and hydrogen bonds.

COVALENT BONDS

A covalent bond is formed by an electron (or electrons) in one atom's outer shell that also spends time orbiting in a second atom's outer shell in order to complete an empty pair. This creates a force that holds the two atoms together: a covalent force.

Covalent bonds result in molecules that have very tightly bound atoms. These are very stable, meaning that they are not usually attracted to other molecules. The ultimate strength of covalent bonds depends on the number of unfilled pairs that are completed—that is, the number

A nitrogen molecule is composed of two nitrogen (N) atoms. As the Lewis diagrams show, when bound together each atom has an octet of electrons in its outer shell, forming a triple covalent bond.

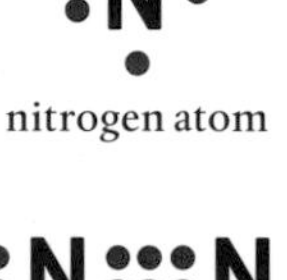

of bonds. Sometimes the bond involves the sharing of only one or two electrons between two atoms; in fewer instances, three electrons are shared. Because more bonds means there is a stronger attraction between the atoms, triple covalent bonds are really, really strong.

The number of electrons available to complete pairs allows chemists to predict chemical behavior. First, they can predict how strong bonds will be. In addition, they can predict how many bonds there are between atoms. (We won't, but chemists use Lewis diagrams to figure this out and make compounds.) The atoms of organic molecules—those made up of carbon and hydrogen and often with the addition of oxygen and nitrogen—are joined by covalent bonds.

There are variations of covalent bonding. A nonpolar covalent bond is one in which the shared electron(s) spend equal time in the orbit of each atom, so that each part of the molecule has the same charge. Because the charge is uniform across the molecule, nonpolar molecules are not attracted to other molecules. Atoms of the same element always bond together by forming nonpolar covalent bonds.

Atoms of different elements, however, have different abilities to share. This results in some interesting behaviors and a second type of covalent bond, the polar covalent bond. Electrons of different elements spend more time circling one atom than the other. The atom that holds the electron longest becomes slightly more negatively charged, whereas the other atom becomes slightly more positive. The molecule ends up with opposite charges at different locations. These locations are called

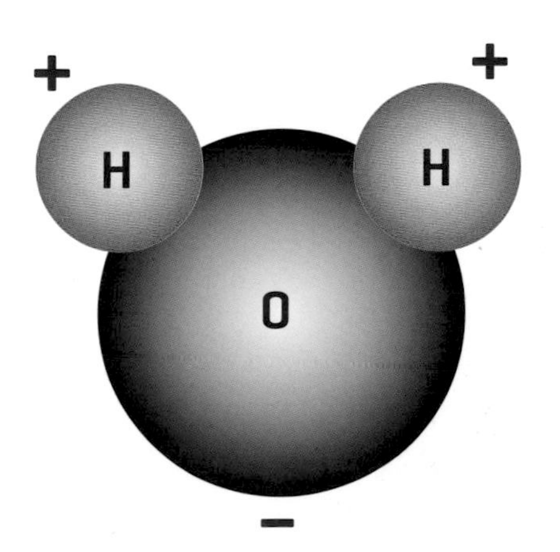

A water molecule is made up of two hydrogen atoms and one oxygen atom. Water is a polar molecule, as it has two different charges on it.

poles. The most famous of the polar molecules is the water molecule, which has a negative charge on the oxygen atom and positive charges on the hydrogen atoms. This is important because on a molecular or atomic level, like charges repel each other and opposite charges attract.

IONIC BONDS

Most atoms are electrically neutral because they contain the same number of orbiting electrons as they have protons in the nucleus. However, for all sorts of reasons, individual atoms can gain or lose electrons and thus develop a charge. A molecule with a charge is an ion. Cations are positively charged ions, and anions are negatively charged ions. Ions are very reactive, meaning they end up combining with other reactive molecules. The mutual attraction of oppositely charged ions results in ionic bonding.

A subscript number in a chemical formula indicates the number of atoms of that element. The presence of a superscript plus ($^+$) or minus ($^-$) sign in a chemical formula indicates that an ion is a cation or an anion, respectively. The number after the charge gives an indication of the magnitude of the charge. Thus, an ion of iron is symbolized by Fe_2^{+2}, meaning it has two atoms of iron (Fe) and two positive charges.

The diameter of an ion's outer orbit determines its size. Hydrogen cations (H^+) are the smallest ions because they don't have any electrons at all. The positive charge comes from the lone proton in the nucleus. For this reason, hydrogen ions are often referred to as protons.

HYDROGEN BONDS

The hydrogen bond is a very special bond, the one that gives water molecules their unusual ability to stick together. Because water is so vital to life, hydrogen bonds have been extensively studied, and discoveries about them are still being made. The field is bewildering, but it all can be distilled for our purposes.

A hydrogen bond forms between a hydrogen atom and a negatively charged atom that is itself part of a molecule or a group of atoms. What is important is that a hydrogen atom can be attracted to two atoms at once instead of just one. The bond usually occurs between hydrogen and oxygen, nitrogen, or fluorine. Hydrogen bonds can form between two molecules or within an individual molecule. They are generally pretty weak, estimated to be about 5 percent the strength of a covalent bond. However, when there are lots of hydrogen bonds, molecules can bond together strongly. This is a great example of strength in numbers.

Most important for us is that water molecules are held together by hydrogen bonds. Water is a polar covalent molecule. The electron sharing arrangement in the molecule results in the oxygen end of the molecule having a higher negative charge, whereas the two hydrogen ends are more positive. When water molecules are close enough, the positively charged hydrogen atom ends form hydrogen bonds with the negatively charged oxygen atoms of other water molecules. So, each water molecule is hydrogen bonded to four other water molecules (and so forth). This is enough to hold them together tightly. Although individual hydrogen bonds are the weakest of the three bonds, when there are lots of them, it takes quite a bit of energy to break them all. Although they break, hydrogen bonds reform continually. This is why water stays in a drop instead of behaving like a powder.

Hydrogen bonds affect the way a molecule is shaped. (Remember, we are thinking in three dimensions.) They cause binding molecules to attach at an angle. It is the presence of hydrogen bonds that results in the famous double-helix shape of the DNA molecule. Hydrogen bonds are

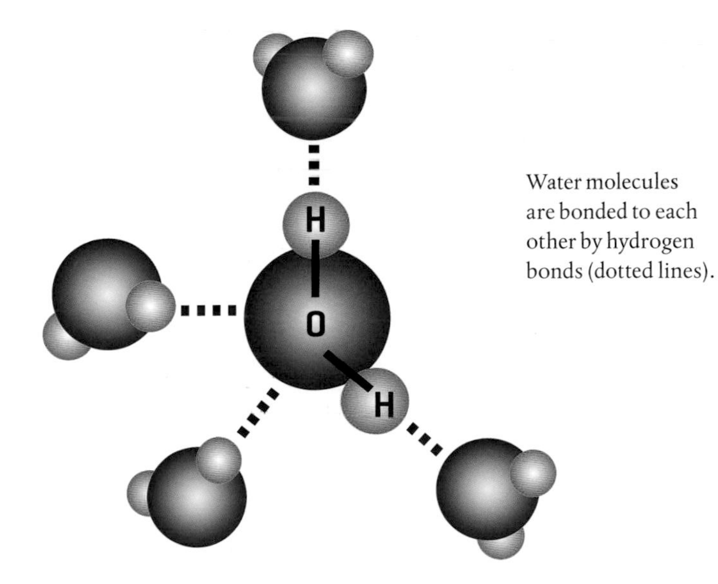

Water molecules are bonded to each other by hydrogen bonds (dotted lines).

also the reason proteins have their various twisted and folded shapes, which make them so functional and important to life processes.

Without getting into complicated chemistry (at this point, at least), hydrogen bonds are why water can exist as a gas, solid, and liquid at the same temperature, and the reason water can be taken up into a sponge or a tall redwood. Hydrogen bonds are why there are smaller temperature extremes near large bodies of water, and why water can remain in a vapor state for so long. All of these have to do with the impact of hydrogen bonds on melting and boiling points, solubility, and viscosity.

Hydrogen bonds explain why water is a great solvent. A solvent is a substance into which other molecules dissolve. Water can dissolve more elements than any other liquid. Charged molecules entering water are surrounded by water molecules and attach to either a positive oxygen atom or a negative hydrogen, depending on their charge. However, hydrogen bonds also hold the water molecule to other water molecules. Thus, the solutes become dissolved in water. In fact, water is such a good solvent that it is extremely difficult to find it in a truly pure state.

Note that in order to be soluble in water, a substance must also be polar or otherwise charged. This is important to how plants take up nutrients because the nutrients are dissolved in water. This means that plant nutrients have to be charged. In almost all instances, nutrients enter plants as ions, some cations and some anions.

ACIDS, BASES, AND SALTS

Chemists use the terms *acid* and *base* often. Gardeners do not. Again, let's put some of our new knowledge to use. Acids are substances that produce hydrogen ions (H^+) in water, whereas bases produce hydroxyl ions (OH^-) in water. Even simpler, an acid is any substance that will react with a base.

Acids and bases are ions. When they react, they form salts. Salts are simply the end products of what is known as a neutralization reaction, two ions neutralizing each other. Salts do not have an electrical charge because the ions that make them neutralize each other.

When put into water, however, salts dissolve and in doing so they split or hydrolyze the water molecule. Salts can create an excess of hydroxyl ions (OH^-) or hydrogen ions (H^+). A liquid's pH is the measure of these ions, so if the salt produces OH^-, it is a basic salt, and if it produces H^+, an acid salt. Many fertilizers are salts. It's a good idea to be able to tell whether things you are putting in the soil are going to make it more acidic or more basic.

THE MOLECULES OF LIFE

Amazingly, plants are simply the combination of four groups of molecules. In fact, so is all life. The only difference between a gardener and a bacterium is the way carbohydrate, lipid, protein, and nucleic acid molecules are mixed.

Carbohydrates are carbon-based molecules made up of combinations of carbon, oxygen, and hydrogen, and for the simpler ones there is usually one water molecule for each carbon atom. The word *carbohydrate* means "carbon water."

Sugars, starches, and cellulose are all carbohydrates. The only difference between them is how many carbon-water molecules they contain. Really simple carbohydrates, monosaccharides (or simple sugars), can have up to six carbons (and that means they can have up to twelve water molecules). Glucose ($C_6H_{12}O_6$) is the most common monosaccharide in plants and animals.

Glucose is used for energy storage as well as a base for adding on

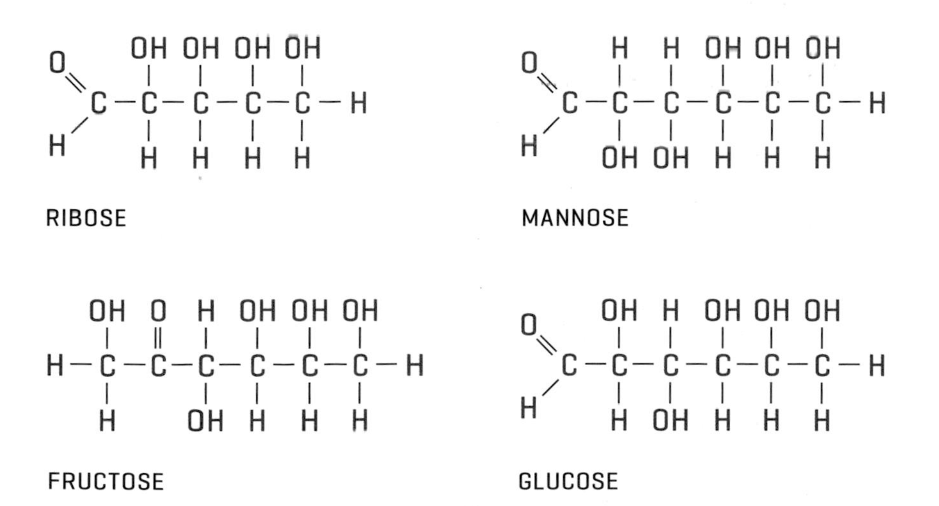

Connecting these simple sugar monomers in various combinations results in more complex and different sugar molecules, known as polysaccharides.

other elements to make lots of other different kinds of molecules. How common is it? The world's plants use more than 100 billion tons of carbon, hydrogen, and oxygen taken from the atmosphere to produce enough glucose molecules through photosynthesis to fill a line of tanker trucks 25 to 30 million miles long.

If you connect two monosaccharides, a disaccharide is formed. Sucrose ($C_{12}H_{22}O_{11}$) is a disaccharide. This is the usual form of sugar when it is transported in a plant. (Pass the maple sucrose, please.) Similarly, if lots of monosaccharides (or disaccharides, for that matter) are linked, they form polysaccharides. Starch $(C_6H_{10}O_5)_n$, the preferred form of sugar for storage in plants, is a polysaccharide. So is cellulose, the material that makes up the outer walls of plant cells. The difference between these two carbohydrates is how the glucose molecules are lined up and linked together. Polysaccharides don't readily dissolve in water.

Lipids include fats, oils, and waxes. The base of these molecules is hydrogen and carbon and a little bit of oxygen. Lipids are nonpolar (no charge, no attraction) and therefore don't dissolve in water. They are hydrophobic, meaning they are pushed together when exposed to water. This is why waxes can block the flow of water. They coagulate in water. It is why an oil-and-vinegar salad dressing has to be shaken to get a semblance of a mixture.

Steroids are also lipids. These molecules consist of four carbon rings to which are attached various molecules that determine their function. Of course, there are the phospholipids, which form all those important cellular membranes that contain transport proteins, and they prevent or delay the entry of unwanted molecules into the cell.

Proteins are nitrogen-based molecules. They are composed of amino acids, each of which has an amino group (NH_2) and a carboxyl group (COOH), plus a side-chain molecule whose composition can vary. (Whew.) The take-home picture is that proteins are large, complicated molecules. They have a lot of bonds that cause them to fold, twist, and change shapes. In fact, proteins are usually folded and the manipulation of these folds results in some of their properties.

Proteins are made up of smaller units, amino acids. Some of these are hydrophilic, and some are hydrophobic. When a protein is exposed to water (and what isn't in life?), it will fold itself up so that the hydrophilic amino acids are exposed to water while protecting the hydrophobic amino acids from it.

Proteins are used to store and move materials, such as membrane transport proteins. Nothing gets through these nitrogen-based molecules unless it's supposed to get through. They can signal to cause the start, end, or change in activities. Proteins create structure and serve as catalysts to speed up a cell's chemical reactions. They are vitally important to every aspect of life.

Nucleic acids, DNA and RNA, are special molecules comprising carbon, hydrogen, oxygen, phosphorus, and nitrogen. These are also very large molecules. DNA and RNA molecules hold the genetic code, the instructions that are used to build cells and everything in them. Among the most important role for nucleic acids is the duplication of proteins, which, despite their numbers and importance, cannot duplicate themselves.

Each nucleic acid is made up of smaller units consisting of a sugar, a phosphate group, and one of two kinds of nitrogen bases. These are nucleotides. Only five nucleotides (adenine, thymine, cytosine, guanine, and uracil) combine in various combinations to make all the DNA and RNA that code life. It is nucleotides, along with hydrogen bonds, that create the famous double helix of DNA.

That is all it takes to build a plant: carbohydrates, lipids, proteins, and nucleic acids. These four primary compounds exist in all living things. All of these organic compounds (molecules containing carbon and hydrogen) fit into one of these four categories, and each and every one of these molecules is composed of the essential nutrients.

OXIDATION AND REDUCTION

The terms *oxidation* and *reduction* are two prime examples of chemists making things more complicated than they need to be. Both are processes that are important to cellular metabolism and occur all the time. In fact, when one occurs, so does the other.

Reduction is simply the addition of electrons, which turns out to be extremely important to cellular metabolism. So is oxidation, which is just the opposite, the loss of electrons. (When scientists say glucose undergoes oxidation during respiration, all they mean is that sugar loses electrons.) Oxidation and reduction occur simultaneously. As the electron is transferred from one molecule to another, one is oxidized and the other is reduced.

ADENOSINE TRIPHOSPHATE

Plants require energy. One molecule in particular, adenosine triphosphate (ATP), serves as the currency of energy in plant cells. ATP, as its name suggests, contains three phosphate ions (PO_4^{-3}). One is attached to a carbon ring that is attached, in turn, to a nitrogen-based molecule, and the other two phosphates are bonded to it. This results in a line of phosphates with two phosphate-to-phosphate bonds.

If a cellular function requires energy, one of the phosphates bonds of ATP is broken to remove a phosphate molecule, thus forming adenine diphosphate (ADP). The energy that held the two phosphates together is released. Similarly, creating phosphate bonds (that is, adding a phosphate to ADP to make ATP) serves as a way of transferring energy for storage.

Phosphorylation is the name given to the transfer of phosphate groups from one molecule to another molecule to release energy. Photosynthesis is all about phosphorylation, as is respiration, which uses the sugar made during photosynthesis. Photosynthesis breaks ATP bonds to make sugar, and respiration uses the sugars to make ATP.

All of the energy used in cells is generated in chloroplasts and mitochondria. In the chloroplasts, chlorophyll captures light energy and uses most of it to convert water (H_2O) and carbon dioxide (CO_2) into glucose ($C_6H_{12}O_6$). Some of this light energy is stored as ATP. Most of

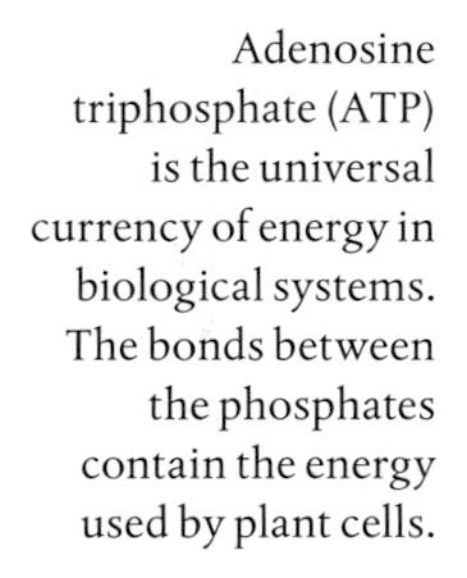
Adenosine triphosphate (ATP) is the universal currency of energy in biological systems. The bonds between the phosphates contain the energy used by plant cells.

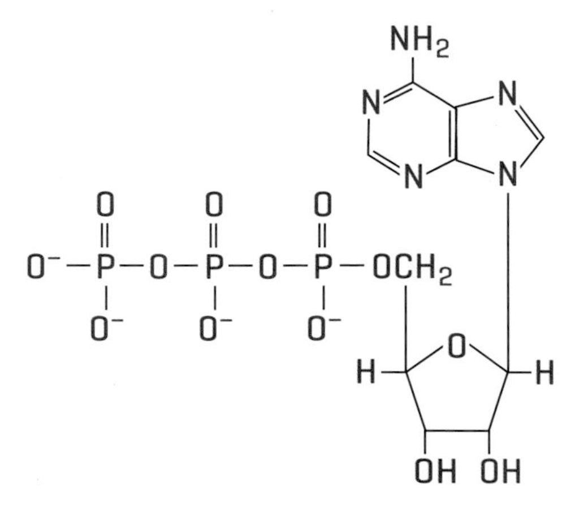

the cellular ATP, however, is produced in mitochondria, where glucose is oxidized to synthesize ATP.

The question arises as to why plants need chloroplasts to make sugar for energy if they can make ATP directly in mitochondria. Part of the answer is that redundancy in key operations is important, but the sugar has other uses as well. We usually don't think of it as such, especially given sugar's bad rap regarding accumulation of both fats and cavities, but plain old glucose is a basic building material for everything in a cell. Its carbon, hydrogen, and oxygen atoms are used to make all of those nucleotides, lipids, and proteins.

In any case, ATP is the energy currency of life. The energy is used to link together molecules and chains of molecules during synthesis. The energy from ATP can make proteins change shape, which serves lots of functions. ATP's energy can break apart water, and it is used to recombine carbon, hydrogen, and all of the nutrients a plant needs.

ENZYMES

Catalysts are substances that increase or decrease the rate of a chemical reaction. Enzymes are proteins that act as catalysts. They work in several ways.

First, enzymes can form a template that holds two or more molecules (substrates) in just the position so that they bind or unbind. Second, an enzyme can get between the atoms of a molecule (or molecules)

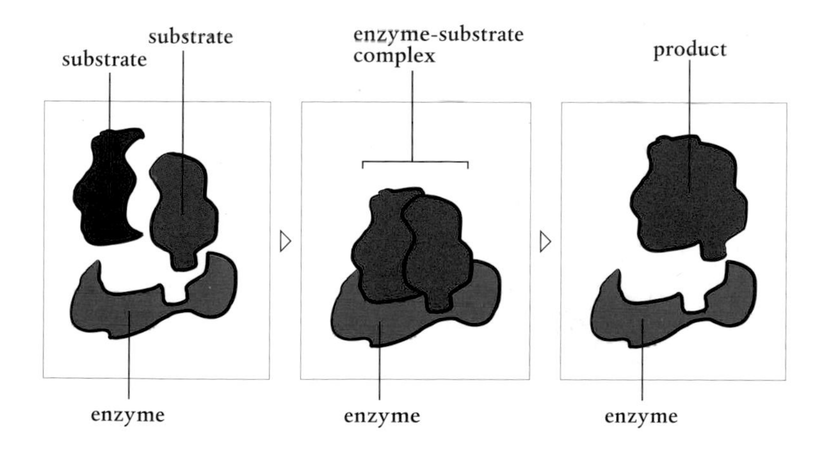

An enzyme in action, connecting two substrates and making a new molecule

and force them into a shape that results in bonding or other reaction. Finally, enzymes can change electrical charges on substrates by taking away electrons (oxidation) or adding them (reduction).

The increased speed afforded by the presence of enzymes is nothing short of phenomenal. Reactions occur millions of times faster, which means nothing is left to chance. It's a good thing these speeds can be achieved in biological systems. Cellular synthesis, decomposition, and a host of other critical activities such as respiration and photosynthesis would simply be to slow to sustain life if it were not for enzymes.

COFACTORS AND COENZYMES

Some enzymes are useless without the presence and involvement of a cofactor, a non-protein molecule that must bond to an enzyme before there is any catalytic activity. The cofactor can't speed things up on its own. Both are needed, which makes cofactors pretty important.

There are different kinds of cofactors, depending on how strong the bond is with the protein enzyme. They can be inorganic or organic, in which case they are known as coenzymes. Many cofactors and coenzymes are vitamins or made from vitamins, and some are metal ions. A metal is an element that conducts both heat and electricity. Several of a plant's trace nutritional elements are metal ions, such as copper, zinc, and iron, which are needed to assist all the catalytic activity in plant cells. They extend the range of enzymes, allowing the same ones to work in different ways.

DIFFUSION, OSMOSIS, AND ACTIVE TRANSPORT

Diffusion is the movement of atoms, ions, or molecules from areas of high concentration to areas of lower concentration. This movement is passive—that is, it does not require the input of energy to the system. For example, if you open a container of really smelly fish fertilizer, the smell soon diffuses throughout the room. Molecules carrying the fishy smell spread out of the bottle, where they are highly concentrated and constantly bumping into each other, and move out into the room, where the concentration is lower. Diffusion normally occurs faster at higher temperatures.

Concentration gradient is the term applied to the difference in concentrations. If everything else is equal, similar molecules always move toward the area of least concentration, or down the concentration gradient, without the application of additional energy. This process is called passive transport.

Osmosis occurs when water is the diffusing agent. If you have pure water on one side of a semi-permeable membrane and water with a solute in it on the other side, the pure water will flow through the membrane to the water that contains the solutes because there is a lower concentration of water molecules in the solution. The solutes, if they can get through the membrane, move in the opposite direction. In this case, it is diffusion, not osmosis.

> Osmosis occurs when water is the diffusing agent

In plant roots, molecules of water, oxygen, ammonium, and carbon dioxide are small enough—or in the case of water have the right chemistry—to move into plant roots via diffusion though the plasmalemma without the input of energy. Some nutrients diffuse into the plant, but instead of diffusing through the membrane, they travel through the proteins embedded in the membrane. These proteins help the nutrients through by changing shapes or utilizing an internal electrical charge, for example, and this is considered facilitated diffusion because it is still diffusion but helped by a protein.

Root cells can buck the rules and get a nutrient molecule to move against its natural concentration gradient. This is known as active transport, and it requires the input of energy. There are several ways to add energy to transport materials against their concentration gradient. Ion pumps, small molecular motors, add energy that can be used directly or indirectly to move ions across a membrane through a transport protein.

ORGANIC MOLECULES

A carbon atom can make four bonds with other atoms, ions, or molecules. Carbon serves as the backbone of all of the organic molecules necessary for life.

Let's consider the implications of being able to make four bonds by starting with atoms that can only make two. Such atoms can only be linked together to form long chains. They can only make linear structures. When an atom has the ability to make three bonds, things take a different shape. Attachments can be made that form three angles and

A carbohydrate is an example of an organic molecule. Technically, organic molecules are those that contain carbon and hydrogen, although some definitions include oxygen as well.

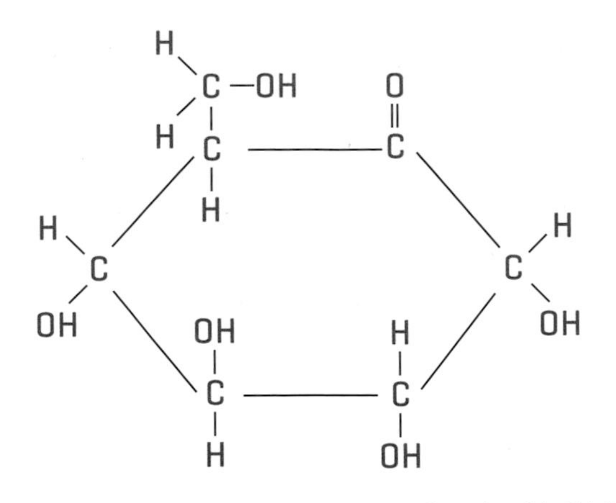

General structure includes monomers all made with CH_2O

three ends on which to make chains and an entirely different shape. You can see where this is going.

When an atom can link to four items, there can be a center with four distinct chains, allowing for much larger, more complex shapes. The backbone of any organic molecule is a carbon atom and its four bonds. The side chains, up to four, give the particular carbon molecule its unique chemical characteristics.

SUMMING UP

That should be enough chemistry to easily follow the rest of the story. Again, don't get hung up memorizing here. Things will be repeated, and you can always come back for reference. If new chemical concepts are needed to flesh things out, they will be explained as we go.

The key things to remember are that chemistry and chemical reactions are all about bonds, and bonds are all about electron orbits, missing electrons, sharing electrons, and transferring electrons.

KEY POINTS

- **Atoms** are the smallest unit of an element. An atom's nucleus contains protons and neutrons. The negatively charged electrons orbit around the nucleus.
- **Molecules** are the smallest unit of a compound. They result from atoms bonding together.
- **A covalent bond** is formed when two atoms share electrons in their outer shells. Molecules formed by covalent bonds are very stable.
- **Ionic bonds** are formed when ions with opposite charges (that is, positively charged cations and negatively charged anions) are attracted to one another.
- **A hydrogen bond** forms between a hydrogen atom and a negatively charged atom that is part of a molecule or a group of atoms. These are generally fairly weak bonds, but they can make up for this weakness by occurring in large numbers.
- **When put into water,** salts dissolve and split some of the water molecules. Salts that create hydroxyl ions (OH^-) are basic salts, and those that produce hydrogen ions (H^+) are acid salts.
- **Carbohydrates** consist of various combinations of carbon, oxygen, and hydrogen. Sugars, starches, and cellulose are carbohydrates.
- **Lipid molecules** have no charge and therefore don't dissolve in water. Fats, oils, and waxes are examples of lipids.
- **Proteins** are large, complicated molecules made up of amino acids. They have a lot of bonds that cause them to fold, twist, and change shape.

- **Nucleic acids** (DNA and RNA) are very large molecules comprising carbon, hydrogen, oxygen, phosphorus, and nitrogen. These molecules hold the genetic code.
- **Oxidation** is the loss of electrons, and reduction is the addition of electrons.
- **Adenosine triphosphate** is the universal currency of energy in biological systems. When bonds between the phosphates in these molecules are broken, energy is released.
- **Enzymes are proteins** that greatly speed up chemical reactions, making them millions of times faster.
- **Cofactors and coenzymes** are molecules that assist enzymes. Vitamins and metal ions are some examples of these.
- **Diffusion** is the movement of atoms, ions, or molecules from areas of high concentration to areas of lower concentration. The movement of water along a concentration gradient is known as osmosis. Both diffusion and osmosis are examples of passive transport.
- **Active transport** occurs when atoms, ions, or molecules are moved against a concentration gradient, which requires the input of energy.
- **The backbone** of an organic molecule is a carbon atom and its four bonds.

THREE Botany for Plant Nutrition

WHAT'S IN THIS CHAPTER

- **Groups of plant cells** are organized into meristematic, vascular, dermal, and ground tissues.
- **Meristematic tissue** is composed of undifferentiated cells that can become any type of cell in the plant.
- **Vascular tissue** is composed of the xylem and the phloem.
- **The xylem** transports water and nutrient ions up from the roots.
- **The phloem** transports water and the sugars and other materials produced by a plant both upward and downward.
- **The dermal tissue** provides protection to the outer surfaces of a plant.
- **Root hairs** are specialized dermal cells that extend into the soil and are the main points of nutrient uptake.
- **Ground tissue** makes up the bulk of the plant body.
- **Leaves** are plant organs that specialize in photosynthesis, which is the production of sugars from carbon dioxide, water, and the energy captured from sunlight.

- **Stomata** are leaf pores that let in carbon dioxide and let out water vapor.
- **The evaporation** of water from the leaf surface, in combination with water's unique chemical properties, is what draws water into the vascular system.
- **Plants form** symbiotic relationships with specialized nitrogen-fixing bacteria and mycorrhizal fungi, and these partnerships are key for the uptake of nutrients by plants.

JUST ONE LOOK at plants and you can tell that they're made up of lots of different kinds of cells, not just the typical one I have been describing. Although plant cells may start out basically the same, at some point in their lives they differentiate and specialize in order to carry out the various functions required to enable the plant to grow and stay alive long enough to reproduce. It is these functions that allow scientists to organize plant cells into four types of tissue: meristematic, ground, vascular, and dermal tissues.

MERISTEMATIC TISSUE

The appropriate place to start is with the meristematic tissue because it is the source of all new cells for plant growth and expansion. Meristematic cells are small and undifferentiated and have the ability to divide, a trait most plant cells lose at maturity. These cells have very large nuclei, small vacuoles (if any at all), and very thin cell walls. They are closely packed together. Meristematic cells are found only in the areas of the plant where growth actually takes place: at the tips of stems aboveground and just above the tips of roots underground. These are known as the shoot apical meristems and the root apical meristems. The development of these primary meristem cells adds to the height of the plant and the length of the roots, but not to the girth. Increases in diameter come from secondary or lateral meristem cells that surround trunks and limbs.

The cytoplasm of a meristematic cell is undeveloped. There are only precursors to plastids, for example, so the cell can be made into one containing chloroplasts or not. It just takes the right signal and enzymes go into action. Meristematic cells have very small vacuoles but lots of protoplasm filling them. This protoplasm is a stew of plant cell building blocks. Once a signal comes in, it provides the elements for construction. It's like having a lumber and hardware store right there on the building site during the construction of a home.

One tenet of the Cell Theory is that cells multiply by dividing. The word *meristem* comes from a Greek word meaning "division." Once a meristematic cell has differentiated, however, it loses the ability to divide. So, meristematic cells start out undifferentiated, and as division produces new cells the older meristematic cells are pushed outward or upward. At this point, a very complex signaling system begins the process that results in differentiation and specialization of the new cells. Thus, the meristematic cell transforms into one that is then part of another tissue group.

During this process an unbelievable number of proteins, many in the form of very specialized enzymes, flow into the cell. Hormones and other plant-made compounds, such as auxins that give plants directional growth, all activate processes that result in a cell's differentiation. The transport proteins in the plasmalemma of meristematic cells are very active, as there is a tremendous amount of activity in the cytoplasm. The large nucleus responds to chemical signals that come though the plasmalemma and begins to direct the construction of ribosomes to process amino acid chains into proteins. The vacuole grows to hold nutrients and to regulate the pH of the cell. It also sequesters the growing amount of waste produced. Energy is provided by the mitochondria, and the brand new transit system of microtubules and actin fibers is put to use and enlarged.

During the period of cellular differentiation, there is a tremendous amount of development. RNA copies the DNA in the nucleus of the cell. Messenger RNA, using instructions copied from the DNA blueprints, constructs all manner of proteins in ribosomes attached to the endoplasmic reticulum as well as floating free in the cytosol. These proteins

MERISTEMATIC CELLS

Meristematic cells are the completely undifferentiated cells in a plant. Upon receipt of a signal, a meristematic cell develops into a specialized cell with a specific function, losing its ability to develop into any other kind of cell. In the aboveground parts of a plant (shoot apical meristem), new cells are constantly formed just below the differentiating ones. Those in roots are located above differentiating cells (root apical meristem).

At the top of this photo is a small group of slowly dividing meristematic cells located in the central zone of the meristem summit in apical meristem tissue. Cells of this zone function as stem cells and are essential for meristem maintenance. A short distance away (middle), the meristematic cells begin to differentiate. In this instance they can be seen developing into leaf epidermal cells that contain stomata (bottom). Had a different signal been sent to the undifferentiated meristematic cells, they would have taken on a different structure and served a totally different function—all still broccoli, but pretty amazing nonetheless.

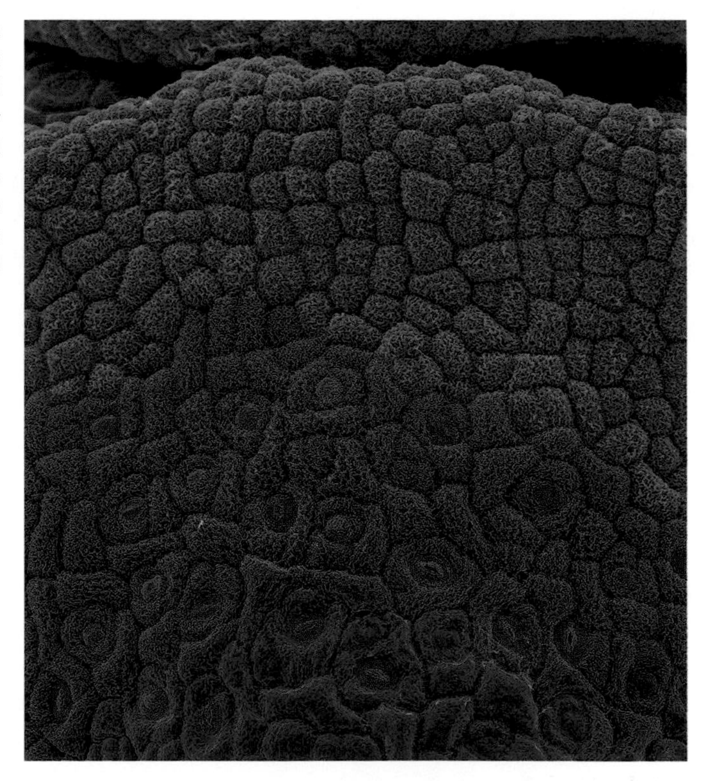

The tip of a small broccoli flower bud with actively dividing apical meristematic tissue along the leading edges of the bud

are transferred out of the endoplasmic reticulum or the cytosol (depending on where the ribosomes are located) to the Golgi apparatus for final packaging. Then the proteins move to the places in the cell where they are needed or to the plasmalemma to serve as a transport protein or to be transported to another part of the plant and combined with some other molecule to make a new compound.

Just what kind of cell a meristematic cell becomes depends on what the plant needs as a whole. All plant cells are connected by their plasma membranes and countless tiny plasmodesmata. One part of the plant can signal a need, and that signal is transmitted to one of the two types of meristematic tissues so cells of that type can be created.

Primary meristematic cells are responsible for producing cells that result in primary growth. This is growth that causes a plant to get taller. For the most part, growth doesn't happen by adding cells, but rather by cells elongating. Primary cells are the cells that elongate.

Primary growth occurs at the apical meristems, which are located just above the root cap and in shoot tips. Hormones signal apical meristem tissue in the stem to produce cells in only one direction so the stem will grow upward. Root meristem tissue, on the other hand, grows in two directions, downward and outward.

Secondary growth is the growth that increases a plant's diameter. This is what produces growth rings in woody plants. Secondary growth happens in lateral meristem tissue, which receives hormonal signals to differentiate. Not all plants undergo secondary growth. Monocots (maize, grasses), for example, lack lateral meristematic cells.

Undifferentiated truly means undifferentiated. For example, a shoot meristem cell can be moved to a growing tip where it will differentiate into a bud, leaf, or flower cell. In fact, each meristematic cell is capable of cloning the entire plant. All of the ground, vascular, and dermal tissues come from meristematic tissue—that is, all of the plant's organs: its leaves, stems, and roots.

GROUND TISSUE

Ground tissue provides most of the mass of a plant as well as its support. These cells also serve as sites for photosynthesis, food storage, protection, and regeneration after injury. There are three types of ground tissue cells, again grouped by function: parenchyma, collenchyma, and schlerenchyma.

Parenchyma cells are responsible for metabolic functions. They have thin primary walls and large vacuoles. In fact, of all the cells in the plant, these are most like the typical cell described in chapter 1. Although parenchyma cells retain the ability to divide, they don't use it very often. This trait allows a plant to regenerate parts of itself when damaged or when it is under extreme stress. Parenchyma cells are not fully specialized cells, and they retain the ability to become more specialized. They are capable of growing, and they function in photosynthesis or storage. There are lots of parenchyma cells in leaves and roots. Fruit flesh is made up of parenchyma cells, and potatoes and other tubers are full of them.

Collenchyma cells form bundles and strands (the strings in celery, for example) and help support plants. Because they lack a secondary (cell) wall and lignin, collenchyma strands are flexible and the plants they support are as well. They can gently bend in the wind, as it were.

Parenchyma cell

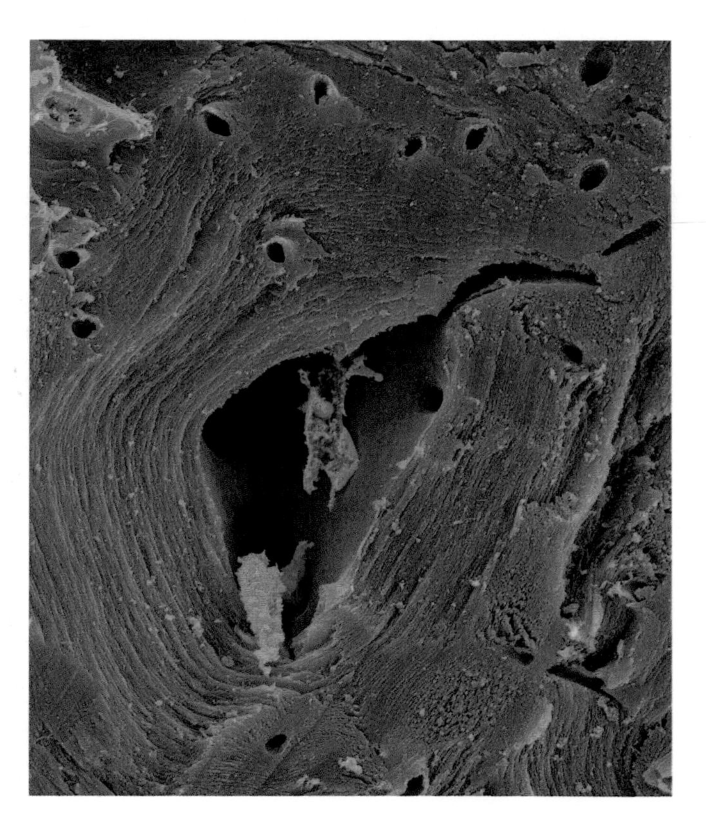

Schlerenchyma cells develop a lignified secondary wall and cannot elongate. These cells serve as support and provide some physical protection to the plant. In fact, most are no longer living. They die when they reach maturity, leaving their lignin shells. Sclereids, a special kind of schlerenchyma cell, make up the hard coatings on seeds and hard nut shells, as well as the skin of a pear, which is why it is harder and tastes different from the skin of apples.

VASCULAR TISSUE

The vascular tissue consists of two types of cells that transport materials in a plant. The xylem carries water and nutrient ions in it, and the phloem transports sugars and organic molecules synthesized inside the plant. Almost no cell is further than a few neighboring cells away from the vascular system.

Phloem and xylem cells are appropriately known as vessels, as they hold water. They run throughout the plant, side by side but separated by the vascular cambium, a wall of lateral meristematic cells that provide

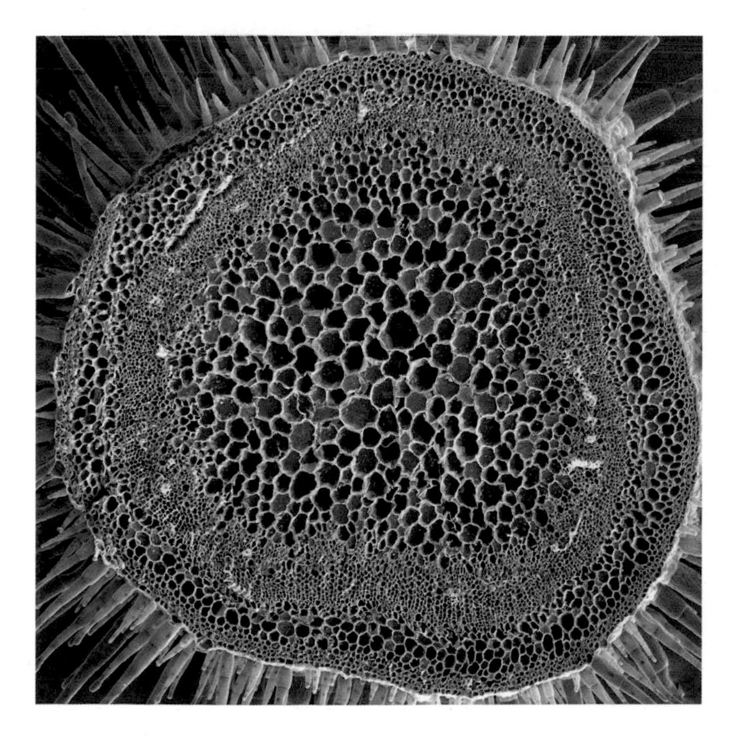

This scanning electron micrograph of a leaf vein in cross section shows the outer epidermal layer (with spiky projections known as trichomes), followed by the cortex and then large vascular bundles, which contain the phloem (nearest the cortex) and xylem.

lateral growth. As the vascular cambium cells differentiate and mature, they develop into phloem and xylem cells.

► **Xylem** The xylem tissue plays a huge role in the story of how plants eat. The xylem is a system of specialized cells, connected end to end, that transports water along with mineral nutrients dissolved in it. This one-way system carries water upward to the leaves of the plant, where xylem tissue is contained in veins. These are placed so efficiently that no single plant cell is further than about 0.5 millimeter (0.02 inch) away from one and so can be readily reached by water traveling from the xylem system and then through cells using the symplastic, apoplastic, and direct pathways. Water in the xylem only flows upward from the roots to the leaves. The connections between xylem cells are much wider than the

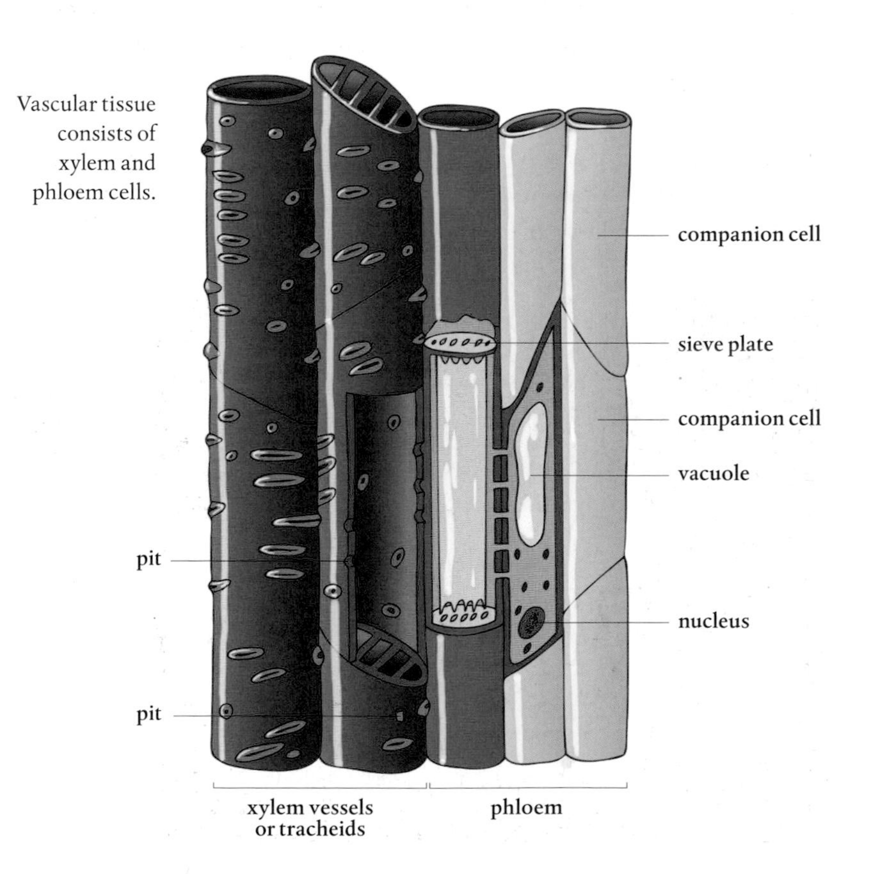

Vascular tissue consists of xylem and phloem cells.

cell-to-cell plasmodesma, thus providing much less resistance to the movement of water.

The xylem's pipes are very elongated cells and are either tracheids or vessel elements that are joined together. Tracheids are the dominant of the two types in gymnosperms, including conifers, cycads, and gink-goes, all softwoods. The end walls of tracheids have pits, thin areas through which water can flow. These cells are tapered and overlap, and water flows from one tracheid cell to the next. They are about the width of a human hair.

Vessel elements, on the other hand, dominate in hardwoods. The ends of these xylem cells are perforated with actual holes. In addition, vessel elements are actually connected, not just overlapping, so there is even less resistance to the movement of water than through tracheid pits. The vessel elements are about six times wider than tracheids, measuring about 500 microns. When linked together, they form vessels that can be a whopping 10 cm long, about the length of your forefinger.

Both tracheids and vessel elements have walls strengthened with lignin, the material that makes wood. This provides support, preventing xylem tubes from collapsing in on themselves. Xylem cells die at maturity, but the lignin allows their shapes and functionality to remain after

Scanning electron micrograph of a conductive vessel element (sieve tube) in wood

their living contents disappear. The lignin also helps to waterproof the tubes to form a continuous, unidirectional plumbing system that moves water and its contents upward from the plant roots through the stems and branches to the leaves.

► **Water Transport in the Xylem** Water transport via xylem is one of the wonders of nature. For starters, it all occurs without the expenditure of any energy by the plant. Imagine how much power would be required to pump water up to the top of a redwood tree. Yet, that tree can bring water to its leaves without using its valuable energy reserves. We would not have tall trees if energy were required to deliver water through them.

Water's journey into the plant starts at its roots. By now you should be aware that water can move into a root cell and travel for a bit in the cell wall via the apoplastic pathway. Once it hits the Casparian barrier, however, water has to move through the cell wall and cell membrane (the plasmalemma), probably via an aquaporin, thus entering the cell and the symplastic pathway. Once inside the plant cell, water moves from cell to cell and makes it way to the xylem tissue located toward the center of the root.

This raises two questions. First, why does the water enter the plant? Molecules of water enter the plant root because of root pressure. There is usually a higher concentration of nutrients than water inside root cells. As a result of osmosis, water outside the roots moves across the root cell membranes to dilute these concentrations inside the cells.

The second question is why water inside a root moves toward the xylem system. Here, there are three forces at play: transpiration (the evaporation of water from the surface of the plant), water cohesion, and water adhesion. To understand transpiration, one needs to consider the leaves of the plant. The stomata are leaf pores that open during the day to let in the carbon dioxide needed to make sugars. At the base of these openings are the ends of xylem tubes. Water in the xylem evaporates out into the atmosphere when the stomata are open. Some leaves also have trichomes, tiny hair-like structures that help draw the water that evaporates out, up, and away from the leaf body where there is more air circulating, speeding the evaporative process.

As we learned in the chapter on basic chemistry, water molecules are polar, with one side having a negative charge and the other a positive charge. This causes water molecules to stick together—that is, to show cohesion. If you lined up water molecules one after another the entire line could be pulled like a rope because of this attractive force. This means that when one water molecule leaves the stomata and evaporates away, it pulls up the molecule behind it, and this happens all the way down the chain. This creates a negative pressure at the bottom of the column, which pulls new water molecules into the system. Water in the roots is literally pulled up through the xylem straw. Once inside the xylem, the water molecules and ions dissolved in it can travel at up to 30 meters (100 feet) per hour.

Ah, but there is more help. The force of adhesion causes water molecules to stick to the surface of their container, in this case the xylem tubes. Adhesion is why a concave meniscus forms when water is held in a glass tube or straw. The water on the edges is higher than the water in the middle of the surface, and because all the water molecules stick to each other there is a tension created. The tension is enough, in combination with water cohesion, to fill up a redwood-sized xylem system.

At night, the stomata receive signals to shut and transpiration ceases. This is the time when plants respire; they utilize the water and nutrients

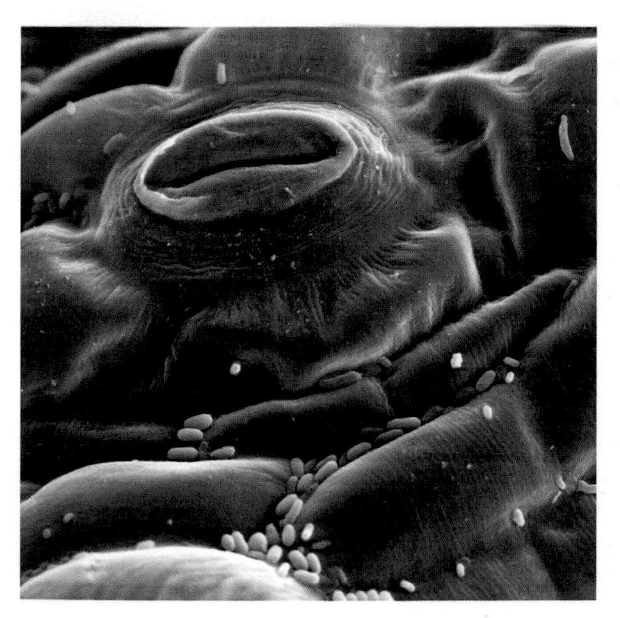

High magnification image of a lower leaf surface showing stomata and some fungi attached to the leaf

brought into the system by transpiration during the day. Although water is no longer moving upward through the xylem, cohesion and adhesion hold the water column in place.

► **Phloem** The phloem is the other half of a plant's vascular tissue. Its function is to transport the sugars made during photosynthesis—along with the amino acids and other organic compounds synthesized by cells, even RNA—to where they are needed. The phloem system consists of only living cells and operates bidirectionally, though not in both directions at the same time.

The phloem distributes sap from a source, such as a chloroplast in a leaf cell or a storage cell in a root, to a sink, with cells in low supply. During summer, the source is usually in the leaves. Sap flows through the phloem to the roots when things are stored for the winter. In the spring, these stored nutrients move from the roots up into the aboveground parts of the plant, only on this trip the sap travels in the xylem. This is the source of maple and birch syrups and is why trees are tapped only in the spring.

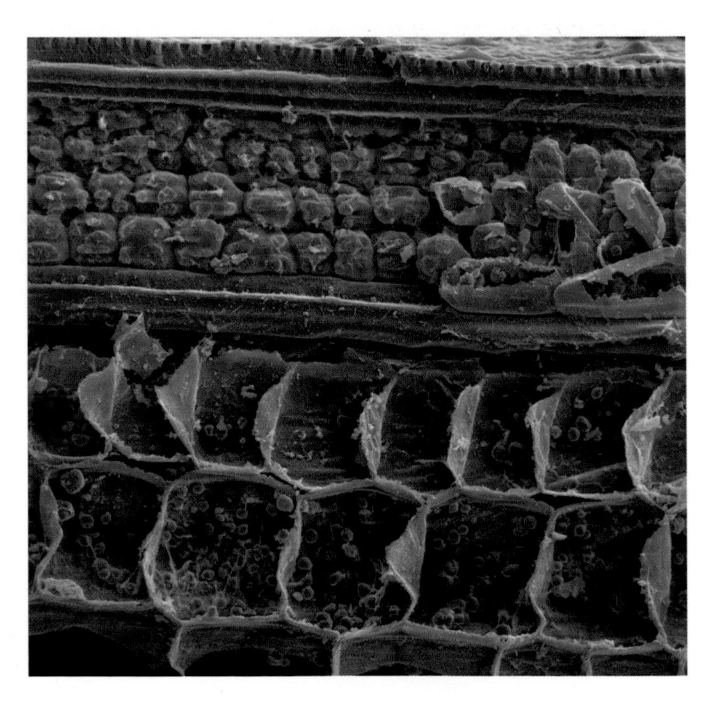

This scanning electron micrograph shows the vascular (bottom layer) and cortex cells of a stem. Starch granules are visible inside vascular cells, and chloroplasts are visible inside cortex cells.

Phloem cells are specialized and elongated cells that lack lignin. These sieve elements are connected in vertical stacks to form sieve tubes, which are separated by sieve plates, perforated structures. The holes of a sieve plate are lined with callose, a polysaccharide. As they mature, sieve element cells lose their nuclei, vacuoles, and ribosomes, but they retain plastids, endoplasmic reticulum, and mitochondria. At maturity, sieve element cells are alive in a way that xylem cells are not.

Because each sieve element is missing a nucleus (and other important organelles), it functions by pairing up with a companion cell that carries out the functions of its missing organelles. Sugar molecules to be transported by the phloem system are dissolved in water and flow into the companion cell, where they are collected and then actively loaded into the sieve tubes. Energy is required to move the sugar into the system.

The phloem depends on hydrostatic pressure, not transpiration, to move the water that contains sugars and synthesized molecules. As sieve tubes are filled with sugar molecules, the concentration goes up inside the tubes. Osmosis causes some of the water molecules in adjacent xylem vessels to flow in to dilute it. There is only so much room for the water moving into the cell, which eventually starts to expand. The positive pressure created inside the phloem pushes the sap through the system.

At the same time, at the other end of the phloem system there is an area of low concentration of ions, the sink, which will receive the phloem sap. Ions flow into the sink through plasmodesmata, to lower the concentration of sap at the source. This relieves some of the pressure in the system, which also causes sap flow toward the sink.

► **Stems** While not part of the vascular tissue, stems evolved to protect it as much as to provide support for the leaves, fruits, and flowers. Specialized cells on the outside of stems protect the insides and act as waterproofing. The stem also has cells that act like Styrofoam packing around the xylem and phloem to protect the vascular tissue.

DERMAL TISSUE

The dermal tissue forms the skin of the plant, keeping in water and protecting the plant from external injury. There are two parts to the dermal tissue: epidermis and periderm. All plants have epidermal cells, whereas only woody plants produce a periderm. Both are composed of tightly packed cells that help provide protection.

The epidermis is usually just a single layer of cells that covers plant surfaces. This tissue is found on flowers, leaves, stems, roots, and even seeds. In general, epidermal cells lack chloroplasts and are transparent.

Epidermal cells produce cutin, a waxy substance that helps form a cuticle that covers all aboveground surfaces of the plant and prevents water loss. The cuticle is usually thicker on top of leaves, where there is more exposure to sunlight, than on the bottom of leaves.

Epidermal cells include the stomata, which contain guard cells that open to allow carbon dioxide to diffuse into the leaves and water vapor and oxygen out. In addition, trichomes, tiny leaf hair cells that help in transpiration, are part of the epidermal tissue. Some trichomes possess glands that produce anti-grazing substances that protect plants from herbivores. Others act as foils, decreasing the flow of air over the surface of the plant and slowing the loss of water from it.

As some plants age, a periderm is formed from lateral meristematic cells. This replaces the epidermis of stems and roots and is more commonly called bark. The periderm not only provides protection, but also allows airflow into the plant. Cork, dead cells full of suberin, is made from periderm.

► **Root Hairs** Specialized epidermal cells on roots, known as root hairs, are extremely important to the uptake of nutrients. Root hairs dramatically increase the surface area of the root that is exposed to soil. Root hairs form because epidermal cells along the outer surface of a root have the ability to change from their normal shape and placement on the surface of a root into an extension of the root that grows out into the soil. Each root hair is a single cell, and these cells can grow to amazing lengths of up to 1500 microns (0.06 inch) and can be as thick as 15 microns (0.0006 inch).

Only a small zone located just below a growing root tip contains meristematic cells coded to become root hair cells. These cells live only from a few days to a couple of weeks, depending on the plant and conditions. Root hairs are constantly being replaced as the growing tip moves into new territory. Still, an astonishing number of these cells can grow at one time. Some plants produce tens of thousands of them in an area no bigger than an inch or two around. For example, studies have shown that a single rye plant can have miles of root hairs growing in 2 cubic feet (0.06 cubic meters) of soil.

Only some of the root epidermal cells form root hairs, because those that do secrete chemicals that prevent their neighboring cells from becoming root hairs, too. If this didn't happen, plants would have huge tangled roots, made up of ever-expanding epidermal cells.

Root hairs act as the eyes of a plant root. Calcium, one of the essential plant nutrients, moves into the root hair through the plasmalemma. The presence of calcium is necessary to complete the process that causes root hair cells to grow downward. As long as there is calcium coming into the root hair from the soil, the root hair cell elongates in the same direction. When an obstacle, say a small rock, is encountered, then that surface of the root hair cell can no longer take up calcium.

Scanning electron micrograph of root hairs developing on a radish during seed germination. These cells release a hormone that prevents neighboring cells from developing into root hairs.

Elongation in that direction stops, but it starts in another area of the cell where calcium does come through the membrane. Once the obstacle has been passed by the root hair, it reorients itself and travels in the original direction. Consider that this goes on all over those miles of hairs. Calcium shortages have obvious consequences here.

Root hairs are tremendously important, but their development is discouraged by the presence of too many soil nutrients. Producing root hairs takes energy and nutrients, and a plant only does this when it is necessary to get more nutrients. If there are enough nutrients, why expend the energy?

LEAVES

The key function of leaves is to provide the platform for photosynthesis. They evolved to best expose as many chloroplasts to sunlight as possible. Leaves also house the stomata, the exchange organs that allow carbon dioxide to move into a plant and water vapor and oxygen to move out.

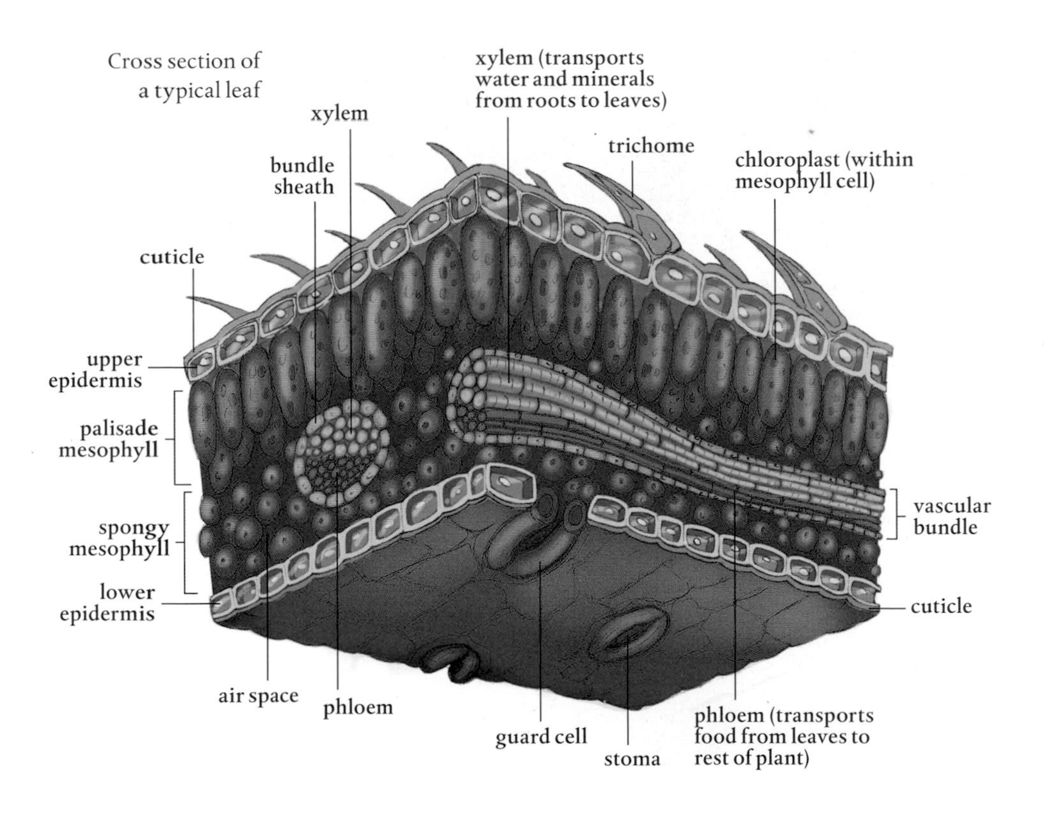

Cross section of a typical leaf

A typical leaf, again assuming there is a typical one, comprises the leaf blade or lamina; the petiole, the stalk that attaches the leaf to the stem; and the leaf axial, a small structure where the petiole attaches to the stem. The xylem and phloem tubes run through the petiole into the lamina. Many plants are identified based on the shape and size of their petioles, which can be circular in diameter, winged, triangular, or flattened.

Leaves are coated on both sides with epidermal cells. Some also have trichomes, hair-like structures, on their upper and lower surfaces. Most leaves are coated on the outside with a waxy substance that forms a cuticle that keeps in water and provides protection. Many metabolic compounds are released into epidermal cells to protect the plant, and these cells are responsible for the regulation of water loss and gas exchange with the atmosphere.

This leaf shows a vein pattern such that no cell is more than three away from a vein.

The chloroplasts reside in between the two layers of epidermal cells, in the mesophyll. The mesophyll is also full of parenchyma and collenchyma cells as well as xylem tubes that provide the water necessary for photosynthesis. The mesophyll also has lots of phloem structures to take away the sugars produced in chloroplasts. When large enough, this network of vascular tissue becomes veins, which are usually sheathed in parenchyma cells and supported by collenchyma cells. The type of leaf venation of a plant is used to describe the plant.

Leaf epidermal cells contain the guard cells that regulate the opening and closing of the stomata. When water rushes into the guard cells, they fill up, swell, and bend on one side and shrink on the other. This results in a hole, the stomatal opening. When guard cells lose water, they return to their flattened shape and the stomata close. Potassium, one of the essential plant nutrients, is key in the operation of stomata.

One function of the epidermis of a leaf is to keep water in when the stomata are closed. However, at night when there is excess water in the soil compared to inside the root cells, a process known as guttation occurs. This is the exudation of liquid water from leaf tips. The water flows into the plant and creates enough pressure to force water into the leaves, where it escapes through hydathodes, tiny holes along leaf tips.

ROOTS

Let's end this brief botany lesson with a look at roots. Of course, there isn't a typical root. Some are fibrous and branched; these fibrous roots usually grow near the surface and are laid out to absorb the maximum amount of water quickly when it rains. Others form the familiar carrot-like taproot that can reach deeper water. Still, there are certain characteristics applicable to most garden plants that make a general model of a root possible.

New cells are produced in the root apical meristem and are pushed out toward the root tip, where they differentiate into columella cells, rectangular cells that line up in columns. These get squashed as new cells are produced above them, and this compaction makes room for new cells. Columella cells form the root cap that protects the apical meristem while the root moves through the soil.

The root cap cells contain statoliths, specialized organelles that appear to help in the detection of gravity and are partially responsible for the downward direction of root growth. These organelles are heavy and sink to the bottom of the cell. In essence, they act as weights and help roots sense gravity. This causes a series of chemical signals to pass through the cell membrane, resulting in increased production of the growth hormone auxin in the area.

Root cap cells help build the mucilage that lubricates the root and the soil, making it easier to penetrate and increasing the contact between the soil and the root. Eventually, the outer cells of root tips are sloughed off and replaced by new root cells produced in the meristematic tissue. These sloughed off cells supplement the plant's exudates (sugars and amino acids, mostly) that attract microbes necessary for nutrient absorption to the rhizosphere. The mixture of microbes living in and digesting this sloughed-off cellular material generates wastes that contain essential plant nutrients, particularly nitrogen.

Mucilage has a great influence on the uptake of nutrients, particularly metal nutrients, into plants. Phosphorus, zinc, iron, and magnesium all diffuse into the mucilage gel and from the gel into the roots. This gel is particularly apt at oozing into soil particle crevices and pores, thus increasing contact with the surfaces where these metals are located. Mucilage contains acids that dissolve phosphorus, which can then diffuse through it to the root. In addition, chemical reactions that occur in the mucilage and the gel's pH result in the uptake of metals. Plant exudates become part of the mixture, produced, at least in part, in direct response to the need for some nutrient or a need to turn off a nutrient supply, which requires a change in the mixture of microbes. In any case, the presence of mucilage contributes to the rhizosphere chemical factory. Some plants take up various metals in greater quantities than others due to the makeup of the root mucilage mixture.

Cells on the upper side of the root's apical meristem move upward. As they do, they form a zone of elongation where, as the name suggests, cells grow by elongating. These cells are little biohydraulic pumps working away to force roots through soil. Their empty vacuoles fill with water as they grow, creating hydraulic pressure inside the cell and making the root tip move downward. Just above this area is a zone of maturation, where meristematic root cells differentiate into their mature form and functions and become part of one of the other plant tissues, the dermal, ground, or vascular tissues.

THE CROSS SECTION OF A TYPICAL ROOT

The root cap protects the root apical meristematic cells. These cells have the ability to differentiate into all of the structures shown in the diagrams: from root hairs, the entry points for nutrients, to the vascular system that transports water up into the plant (xylem) and sugars and other compounds throughout it (phloem). In fact, if you moved a root meristematic cell into a shoot apical position, it would generate into a tip-related part of the plant and not a root-related one.

The elongation of cells in the root provides some of the hydraulic pressure that drives the root forward through soil. Mature parts of the root usually stop producing root hairs. Cells of the root cap are constantly being replaced along with other root cells, and those that are sloughed off add to the carbon that attracts bacteria and fungi to the rhizosphere, the area immediately surrounding roots.

The endodermis is the dividing point between the apoplastic and symplastic pathways for water. It contains the Casparian strip, a layer of cells filled with the waxy substance suberin, which stops the movement of water through the cell walls so that it must travel the symplastic pathway. Because cell membranes are much more selective than porous cell walls, this allows the plant to better control what it takes in.

Different kinds of plants have different vascular bundles, and their arrangement is a distinguishing factor between monocot and dicot plants. All of this specialization starts with the undifferentiated meristematic cells.

If we were to look at a cross section of a root, at the very outer edge we would find the epidermis, or outer skin. This layer of cells covers the entire root except for the area immediately around the root cap. The portion of the epidermis just below the zone of elongation is where root meristematic cells specialize and become root hairs.

The next layer in is the cortex, which is composed of lots and lots of

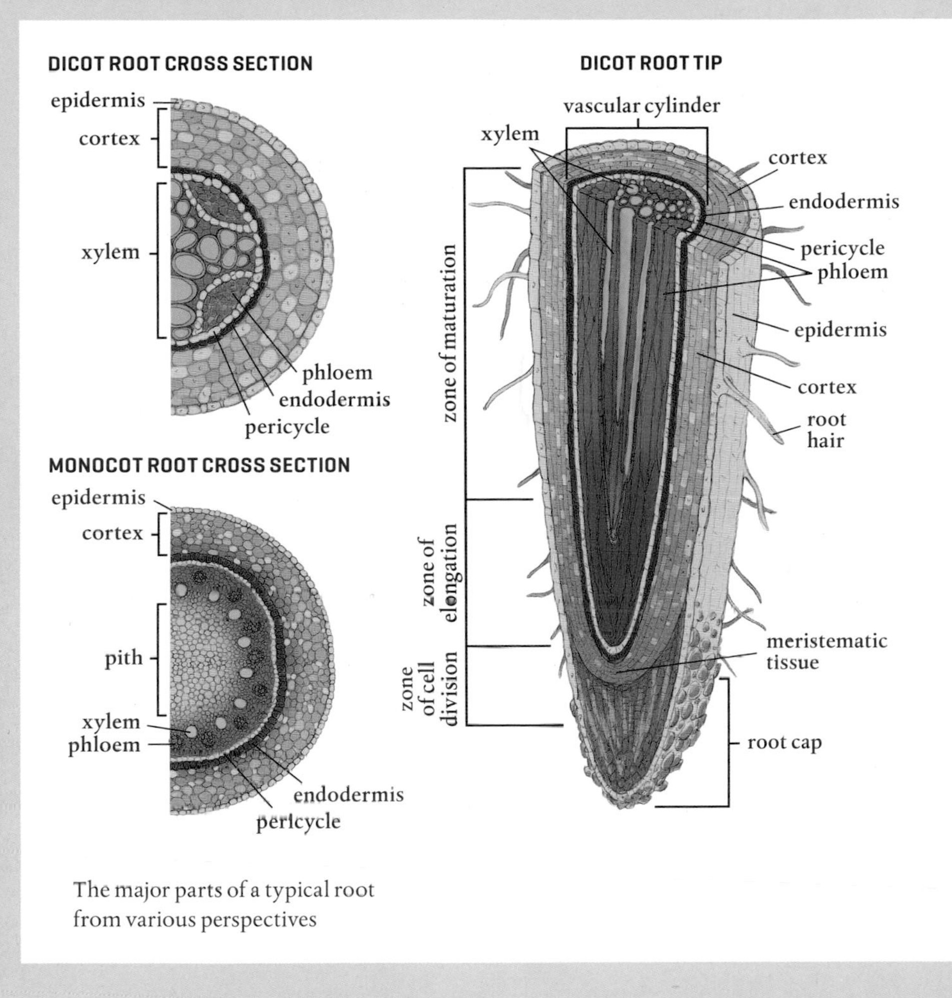

The major parts of a typical root from various perspectives

parenchyma cells that store food, especially in the form of starch. On the inside of the cortex, toward the center of the root, is a layer of cells known as the endodermis. These cells are squished together so tightly that they lack any space between them. This is the location of the Casparian strip, the barrier that prevents water (or anything else, for that matter) from traveling unregulated farther into the plant.

On the inside of the endodermis is the stele, the central part of our model root. The outer boundary of the stele is the pericycle, a thin layer of cells that retain the ability to divide. These form lateral (branch or side) roots, which are quite different from root hairs. They grow through the cortex and the epidermis and out into the soil, often developing root hairs themselves. Pericycle cells form new cells that help thicken the root.

At the center of the root, inside the pericycle, are the vascular tissues. Here the xylem system begins to transport the water that enters the root hairs upward though the plant. The phloem also starts here and brings nutrients down from the leaves to the roots for use or for storage, or it carries stored materials back up into the aboveground parts of the plant.

SUMMING UP

All gardeners appreciate what special organisms plants are, although many never fully appreciate how their parts work. At this point, you should be able to contemplate water and nutrients entering the root hairs of a plant as a result of diffusion, and then moving through the layers of the root and entering the xylem, where these materials move up through the stem and petioles and into the leaves. Similarly, you can imagine the phloem in the root as it moves stored sugars and organic molecules from the storage cells in the root up into the leaves or as it carries these materials down from the leaves where they were produced.

Don't get hung up on all the names used to describe the parts of a plant. Just try and revel for a moment (in three dimensions) over the elegant organization of a plant. Hopefully, I have given you something to contemplate during some of the quieter moments out in the garden.

- **Apical meristematic tissue** is located just below the tips of shoots and just above the root caps. These cells can mature to become any type of cell in the entire plant and are responsible for increasing plant height and length. Lateral meristem cells add diameter to plants.
- **Ground tissue** provides most of the bulk of a plant as well as its support. These cells

also serve as sites for photosynthesis, food storage, protection, and regeneration after injury.

- **Vascular tissue** is made up of xylem and phloem vessels that transport water and nutrients.
- **The xylem** carries water and dissolved nutrients upward from the roots. Transpiration, cohesion, and adhesion allow water to move through the xylem without the plant expending any energy.
- **Phloem vessels** move sap (water and materials produced by the plant) both upward to the leaves and downward to the roots.
- **Dermal tissue** forms the skin of the plant, which keeps in water and helps to protect the plant from external injury.
- **Stomata are pores** in the epidermis that open to allow carbon dioxide to diffuse into the leaves and water vapor and oxygen out.
- **The key function** of leaves is to provide the platform for photosynthesis. Chloroplasts reside in cells between the two layers of epidermal cells that cover each leaf.
- **Root hairs** are extremely important to the uptake of nutrients. Each is a single epidermal cell that can grow to an amazing length.
- **Roots take in** water and nutrient ions with the aid of microbes that live in the soil.
- **Mucilage** is the mixture of sloughed off root tip cells, exudates from root tips, and microbial populations and by-products. It acts as a lubricant for roots growing into soil and influences the uptake of nutrients, particularly metal ions.

FOUR The Nutrients

WHAT'S IN THIS CHAPTER

- **Macronutrients** are those elements needed by plants in the greatest quantities.
- **Micronutrients** are elements needed in only trace amounts.
- **Carbon, hydrogen, and oxygen,** which are obtained as carbon dioxide and water, make up about 96 percent of the mass of a plant.
- **The macronutrients** obtained in mineral form are nitrogen, phosphorus, potassium, calcium, magnesium, and sulfur.
- **All micronutrients** are obtained from the soil, and these are boron, chlorine, copper, iron, manganese, zinc, molybdenum, and nickel.
- **Because nutrients** enter a plant as ions, they have different degrees of mobility within the plant.
- **Where the symptoms** of nutrient deficiencies occur, such as wilted growth tips or yellowed older leaves, depends, in part, on the mobility of the lacking nutrient.

NO MATTER WHERE a plant grows, no matter how complex its flowers or fruits, no matter what its seeds look like or what kind of leaves it grows, all it takes for that plant to survive and reproduce are a mere seventeen of the ninety naturally occurring elements. (There are twenty-seven or so other elements, but these are man-made and thus could not be required for plants to survive.)

Just seventeen elements sustain life—and not just plant life, but yours and mine, as well. We eat plants and/or animals that eat plants to survive. Combining these seventeen elements results in all of a plant's beauty, its physical structure, organs, and ability to sustain life on Earth. This is astounding.

Plant scientists have defined these seventeen elements as the essential nutrients. They are all needed for plants to survive (that is, to grow and reproduce). No other element can replace an essential nutrient because none can carry out its functions. Are there other essential nutrients besides the seventeen discussed here? Probably, but their exact functions have not yet been determined. Researchers are conducting hydroponics studies in which an element is left out of the nutrient solution to test its effects on plant health and growth. This research will likely identify new essential elements, but their necessary quantities are so small that the search is no doubt a difficult one.

A plant may contain other elements in addition to the essential elements, and some of these actually benefit the plant. There may be between thirty and sixty nonessential elements in a plant. Seaweed is famous for containing sixty natural elements. Some nonessential elements play positive roles and improve the plant in some way, but at the end of the day, the plant can survive without them. However, a plant cannot grow and reproduce without each of the seventeen essentials.

Here, again, is something to contemplate while gardening or when looking at a tree or following the progress of an amaryllis as it goes from bulb to flower in mere weeks. The combination of a mere seventeen substances makes all you are looking at. You can't even play a decent card game with seventeen cards, but you can build plants by combining seventeen elements. Do I have your attention?

ESSENTIAL NUTRIENTS

Most gardeners can name many of the essential nutrients. The macronutrients are the ones required in the greatest quantities. Three of these are always represented on fertilizer packages as the N–P–K trilogy: nitrogen (N), phosphorus (P), and potassium (K). Because it often comes up, the symbol K is used for potassium not because the letter P was already taken by phosphorus, but because it comes from the Latin name *kalium*. Beyond this trilogy, some gardeners are familiar with other macronutrients, such as sulfur, calcium, and magnesium. These are also used by plants in large amounts. Carbon, hydrogen, and oxygen are also macronutrients.

The second category is micronutrients, which are sometimes called trace minerals. The lack of iron, manganese, zinc, copper, molybdenum, boron, chlorine, or nickel can cause plants to do poorly. Although the name micronutrient might suggest they are less important than the macronutrients, they have the same degree of importance. They are essential, but only tiny amounts are required. The micronutrients are present in most soils and don't have to be added very often unless there is something way off balance. But they have to be there or the plant will not survive and reproduce.

The list of essential plant nutrients is not a very long one, and most should be familiar to you because they are in your own daily diet (just check your vitamin and mineral supplements bottle). If you're going to be a really good gardener, though, you need to really understand a lot more about them.

The Essential Macronutrients and Micronutrients

MACRONUTRIENTS	MICRONUTRIENTS
carbon (C)	boron (B)
hydrogen (H)	chlorine (Cl)
oxygen (O)	copper (Cu)
nitrogen (N)	iron (Fe)
phosphorus (P)	manganese (Mn)
potassium (K)	zinc (Zn)
calcium (Ca)	molybdenum (Mo)
magnesium (Mg)	nickel (Ni)
sulfur (S)	

MACRONUTRIENTS

Some of the macronutrients are familiar to gardeners in general terms, and some are even familiar in specific ways. Most gardeners, for example, associate a yellowing lawn with a lack of nitrogen. Understanding what each of these nutrients does in cellular terms, however, is the best way for a gardener to be able to assess and address problems in plants, should they arise.

► **Hydrogen, Oxygen, and Carbon** Hydrogen, oxygen, and carbon account for a whopping 96 percent of the mass of a plant. Carbon and oxygen each make up around 45 percent and hydrogen 6 percent. That leaves only 4 percent for the other fourteen essential elements.

Water (H_2O) and carbon dioxide (CO_2) are the sources of a plant's hydrogen, oxygen, and carbon. Carbon dioxide and oxygen can be dissolved in water and enter plants via the roots. However, these gasses mostly enter through stomata on the leaves. Water is the source of hydrogen, and the breakdown of water results in the release of oxygen as a by-product. The carbon and oxygen used in photosynthesis comes from carbon dioxide. Individual plant cells have direct access to air and water at all times due to the unique structure and characteristic of cell walls, which surround every plant cell. You know it as the apoplastic pathway.

Despite their overwhelming presence in the makeup of plants, however, these elements are not considered fertilizer. This is not to say plants don't need water to survive or that a gardener can't supply it. In an enclosed situation, carbon dioxide can be pumped in to help plants grow, but this is not the norm. Hydrogen, oxygen, and carbon are clearly essential, but they are non-mineral nutrients, not fertilizers, and so they are only tangentially covered in this book.

► **Nitrogen** Nitrogen (N) is crucial to plant growth. One could argue why it is so important, but I settle on its presence as the backbone of amino acids, the structural building blocks of proteins, one of the four kinds of molecules that make up life. No nitrogen, no proteins.

Enzymes are proteins. These catalysts are required for all activities in a cell. Nothing happens on a cellular level without an enzyme being involved. And when something doesn't happen, it's usually because an enzyme is missing. Photosynthesis and respiration absolutely require nitrogen and the enzymes necessary to drive these processes. Proteins are also the molecules that make cellular membranes semi-permeable. They are the channels, carriers, and motors that are necessary for sufficient quantities of water and any quantity of the other essential nutrients to get into a plant cell.

Some would argue that a more important role for nitrogen is its role as the base element for nucleotide molecules. These are the building blocks of DNA and RNA, the blueprints and translators, respectively, of the genetic code. Much of a plant's cellular activity, including the production and use of DNA and RNA, is to ensure there is an adequate supply of those nitrogen-based enzymes to carry out cellular activities.

If that isn't enough, nitrogen is also an essential part of the chlorophyll molecule ($C_{55}H_{72}MgN_4O_5$). Without those four nitrogen atoms, there is no photosynthesis. Therein, incidentally, lies the answer to the yellowing lawn: a lack of nitrogen means there is less of chlorophyll's green pigment.

Nitrogen remains mobile once inside a plant, meaning it can be transported to where it is needed. This mobility is also why the first signs of yellowing from a lack of nitrogen occur in older leaves. Nitrogen is so critical to new growth that the plant will rob nitrogen from older cells in order to grow new ones.

Outside of the plant itself, nitrogen also has a great influence on the pH of the soil, which has a direct influence on the uptake of all nutrients. The pH in the rhizosphere goes up when NO_3^- is added because hydroxyl ions (OH^-) are released. This increases the solubility of iron and aluminum phosphates.

Most plants in a garden are about 3 to 4 percent nitrogen by weight. By way of comparison, your body is around 3 percent nitrogen, which makes sense because we really are what we eat. Only carbon, hydrogen, and oxygen exist in higher concentrations in plants (and humans).

The Earth's atmosphere consists of a whopping 78 percent nitrogen. Unfortunately, atmospheric nitrogen (N_2) is off limits to plants because nitrogen atoms form extremely strong, triple covalent bonds with each other. However, it is precisely this ability of nitrogen to form triple bonds (due to empty pairs of electrons in its valence orbit, remember) that makes it so chemically important.

Those triple bonds are incredibly difficult to break apart. Until the early 1900s, when chemists solved this puzzle, atmospheric nitrogen bonds could only be broken by biological means via bacteria and Archaea. Making nitrogen useable by plants is nitrogen fixation, and the soil microbes responsible for this are known as diazotrophs. So important is nitrogen that plants will work with other organisms to get an adequate supply. Up to 50 percent of the nitrogen in your garden can come from nitrogen-fixing organisms.

The most familiar diazotrophs are *Rhizobia*, a group of soil bacteria that form symbiotic relationships with legumes. *Rhizobia* provide the enzymes necessary to break apart triple-bonded atmospheric nitrogen molecules, and the plant provides the housing for this activity, as well as carbon-based exudates that the bacteria consume. The amount of oxygen has to be limited for the enzymes to work, and the root nodules provide such an environment. There is enough nitrogen for the legume and the bacteria to share. Up to 20 percent excess nitrogen is produced as well, and it moves into the soil and soil food web, where much of it is brought to plants by another group of symbionts, mycorrhizal fungi. Many commercial mixtures of *Rhizobia* are available, and we now know that there has to be a specific match between the right species of *Rhizobia* with the right kind of plant.

Less is known about *Frankia*, the so-called filamentous bacteria, or more properly actinomycetes. These nitrogen-fixing bacteria associate with actinorhizal plants that include alders, several trees and shrubs found in the South Pacific, and some more familiar plants used as

THE NITROGEN CYCLE

Atmospheric nitrogen (N_2) is so tightly bound that it is unavailable for use in most biological systems. Plants, however, need nitrogen in large quantities. A tiny fraction is made available by electrical and snow storms, but until the advent of artificial manures in the early 1900s it was the job of specialized microorganisms to break these bonds and make nitrogen available to plants. Some of these nitrogen-fixing bacteria live in a symbiotic relationship with plants in root nodules—*Rhizobia* nodules on legumes, for example. Others, such as *Azotobacter*, are free-living soil organisms. Both have the ability to break the triple covalent bond that holds the two nitrogen atoms together. Today, the Haber-Bosch process allows for artificial nitrogen fixation and only about 50 to 60 percent of available nitrogen is produced by biological systems.

Fixed nitrogen in the form of ammonium (NH_4^+) and nitrate (NO_3^-) is useful as a nutrient to plants and is taken up and assimilated into various compounds within the plant. When plants die, either a natural death or as a result of being eaten by animals, they decompose. The decay by bacteria and fungi results in ammonification, the production of NH_4^+. Some of this is taken up by plants, and the rest is converted by nitrifying bacteria to nitrite (NO_2^-), which is then converted by bacteria and Archaea into NO_3^-. Some nitrate is absorbed by plants and assimilated, some remains in the soil, and the rest washes out of the soil and is suspended in water.

When soil or water that contains nitrogen becomes anaerobic (that is, they lack oxygen), nitrates are converted back to atmospheric nitrogen, as some anaerobic microbes use nitrogen compounds as their energy source. This nitrification is a big loss for the gardener, but it does complete the nitrogen cycle.

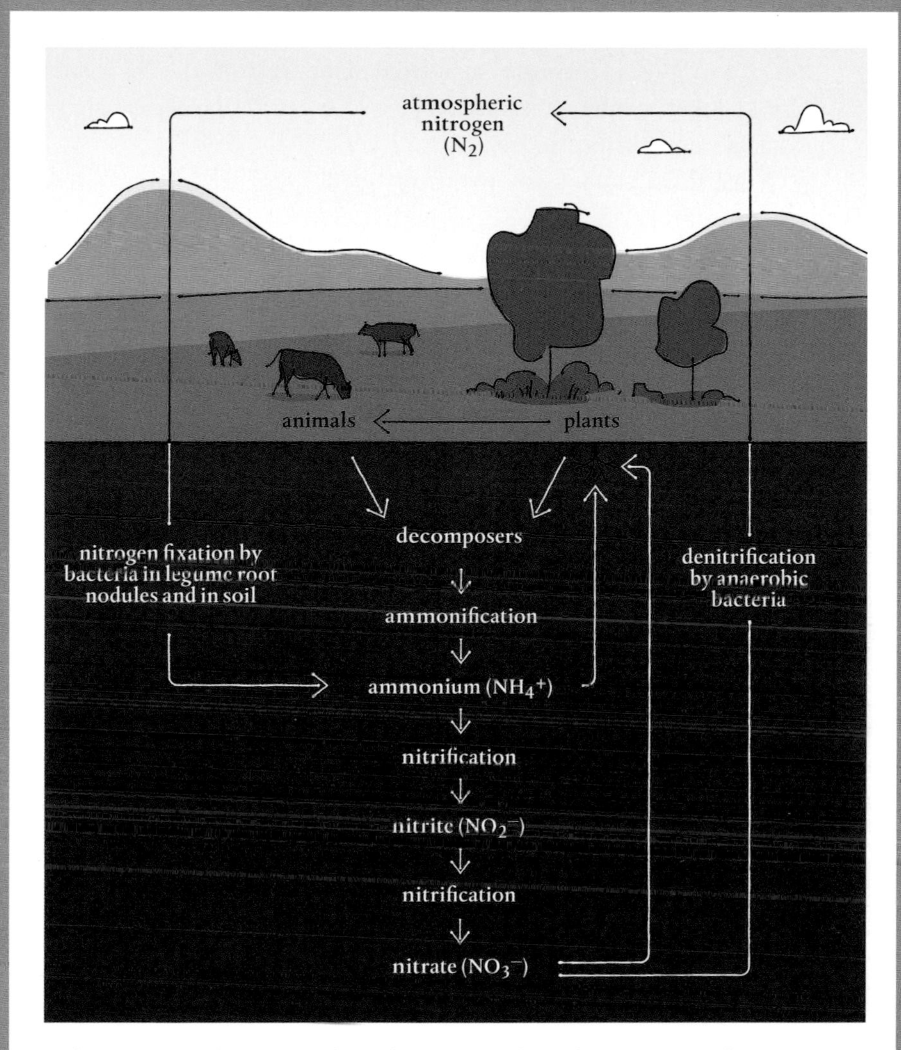

In the nitrogen cycle, nitrogen from the air is transformed into compounds that can be used by plants and is then released back into the atmosphere.

landscape material, such as oleaster, bayberry, dryas, and the edible sea buckthorn. Most of these are considered pioneer plants, alder in particular, and their symbiotic relationships with *Frankia* bring needed plant-available nitrogen into poor soils.

Aside from biologically fixed nitrogen, there are relatively few natural sources of nitrogen. Those that exist are primarily deposits of nitrogen-based minerals and guano, both of which were severely exploited in the 1800s, actually resulting in wars; the rise of the modern German, U.S., and British navies; and the start of modern nation building. Fears that the world would run out of useable nitrogen to supplement the biologically fixed supply had developed, and it appeared that the world would not be able to conduct enough agriculture to sustain a growing population.

Around the start of the twentieth century, the German chemist Fritz Haber figured out how to fix nitrogen. Carl Bosch, an industrial engineer, scaled up the process so it could be accomplished on an industrial scale, and the rest is history. Today the Haber-Bosch process provides over half a billion tons of artificial manures each year, requiring a staggering 5 percent of the world's natural gas production to do so. This artificially fixed nitrogen sustains more than a third of the world's food production.

► **Phosphorus** Phosphorus (P) is extremely important for many of the same reasons as is nitrogen. For starters, it is also a component of DNA and RNA, the molecules that store the genetic code and translate it. A more unique role for phosphorus, however, is as a base for the adenosine triphosphate (ATP) molecule. There are two bonds between the three phosphorus atoms. The bonds contain lots of energy, which is released when they are broken. Plant cells have enzymes that split off phosphorus from ATP molecules and rebind them onto its precursor molecule, adenosine diphosphate (ADP). Breaking these phosphorus bonds produces energy, and making these bonds stores it.

It is no wonder, then, that stunted growth is a sign of phosphorus deficiency. Without phosphorus, there isn't energy to sustain growth. If the deficiency occurs when the plant is older and most growth is finished, then blooming and continued root growth are affected. No new cells can be made at either type of apical meristem.

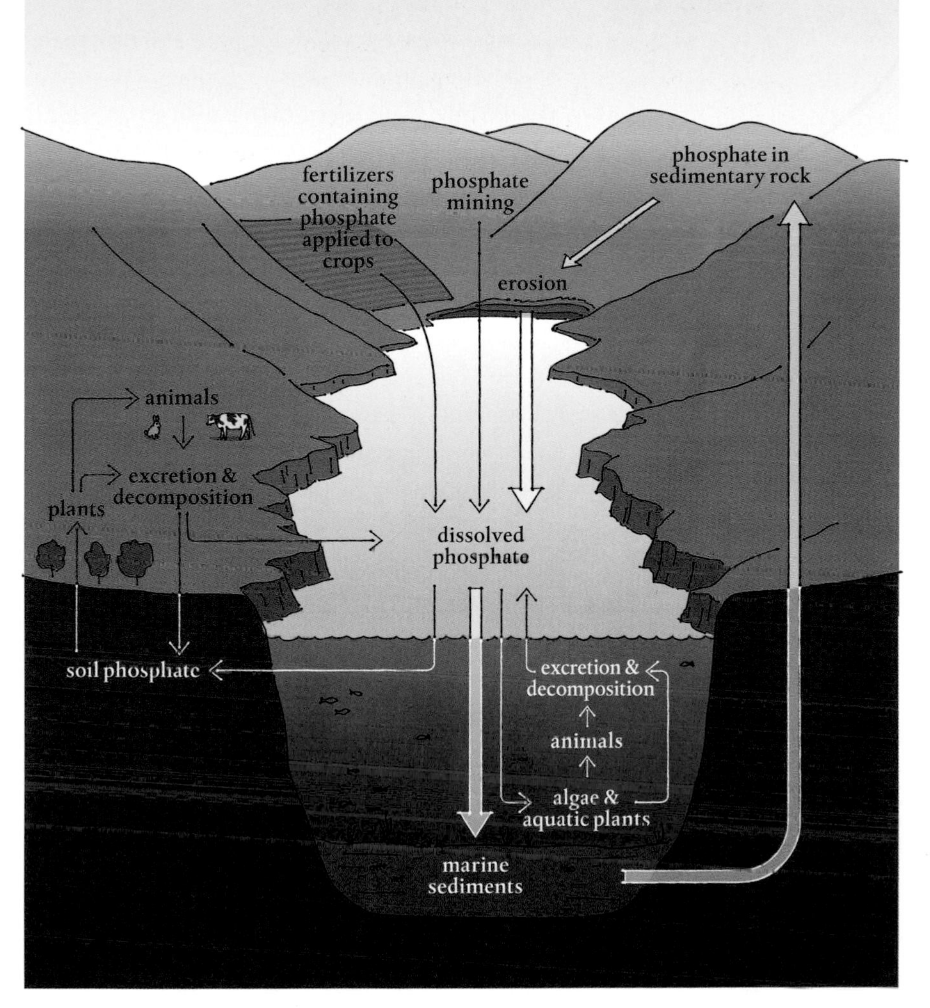

The phosphorus cycle

The list of plant cell functions requiring energy is a long one, but at the top is the need for adequate energy to enable plants to utilize the sugars they produce. When supplies of phosphorus are low, plant leaves turn bluish green. This is caused by an accumulation of sugars, which can't be used because of a lack of energy. As with nitrogen deficiency, symptoms appear first in older leaves, indicating that phosphorus is also mobile in plants and is moved to where it is needed most.

When supplies of phosphorus are low, plant leaves turn bluish green

DNA, RNA, and ATP? Who needs more to convince them of the key role of phosphorus in plants? But there is more. Phosphorus is a major component of the cell membrane, as part of the phospholipids. As such, phosphorus is a key element in the cellular gate-keeping system. It is absolutely necessary to build and maintain the integrity (not to mention the contents) of a plant cell.

Phosphorus ultimately comes from the weathering of apatite, a type of rock. It can enter soils in two forms: organic and inorganic. As you would expect, the former is from decaying dead plants, manures, and microbes that contain it. Like nitrogen, phosphorus also cycles through the environment in various forms. Inorganic phosphorus, from apatite, is adsorbed to the surfaces of clay particles and organic matter. The attraction is an ionic one, binding the phosphorus in place, and as a result it does not move much in soil. In fact, it is so tied up that 1 acre (0.40 hectare) of soil might have only 1 pound (0.45 kg) of phosphorus dissolved in water and available to plants. In any case, eventually phosphorus makes its way back into the ocean, where it is redeposited and again becomes apatite rock—although it takes a heck of a long time.

In fact, because phosphorus becomes tied up in the soil, tremendous amounts are applied to agricultural fields (and, too often, gardens). Many believe that we have squeezed the phosphorus cycle to the point that the world is going to reach the peak of production and then quickly run out of phosphorus. More than 22 million tons (20 million tonnes) of it are being applied to fields every year. The Global Phosphorus Research Initiative estimates that mined supplies will be insufficient for crops in only about 30 or 40 years. Because 95 percent of the remaining mineable phosphorus is found in Morocco, China, South Africa, Jordan, and the United States, it is possible that this nutrient will gain even more essentiality, possibly even causing intense military as well as economic competition.

Because of the tight adsorption to soil particles, phosphate uptake requires that roots grow so they can maintain continuous and new contact with phosphorus. In addition, around 95 percent of all plants associate with mycorrhizal fungi, which provide the plant with phosphorus in return for the carbon in root exudates.

► **Potassium** Potassium (K) is the only essential nutrient that is not a constituent of any organ, organelle, or structural part of plants. Its role is as a regulating chemical. As such, potassium is the key solute in the cell's cytosol, where it can exist in high concentrations without causing damage. Potassium ions play a key role in the movement of water into and out of cells. Those all-important guard cells in leaves open and close as a result of different potassium concentrations. This is how the plant as a whole regulates carbon dioxide and water levels.

A potassium ion (K^+) is positively charged, making it a cation. Potassium's positive electrical charge acts as a counter balance to other charged molecules in plants. Potassium also regulates more than sixty key enzymatic reactions, speeding up chemical reactions by thousands of times. Its presence is crucial for the formation of starch, which is used to store the sugar made during photosynthesis, and for the movement of sugars themselves.

Potassium is mobile in plants, and older leaves show the signs of deficiency, first in the form of dead spots due to dead cells. If the mineral or water balance cannot be regulated, cells will die. Plants low in potassium will often wilt because of its role in regulating stomata. If stomata don't operate properly, water escapes, with disastrous consequences.

Potassium is held by negative charges on the surfaces of clay particles and organic matter in soil. It is not adsorbed tightly, however, and is available for plant uptake. As far as sources, potassium is the eighth most abundant element, making up about 2.5 percent of the Earth's crust. It is found in all sorts of minerals.

► **Calcium** Calcium (Ca) is a key structural component of cell walls. Because these form around every plant cell, this element is of immediate and obvious importance. Calcium is also a part of many enzymes, and it is important as a signaling molecule. Under normal cell conditions, the cytosol has a very low calcium concentration, with supplies sequestered in various organelles, such as the vacuole or endoplasmic reticulum. Increasing concentrations of calcium activate enzymes in the cytosol, speeding up reactions and causing things to happen.

Calcium is also used to transport other substances across the cellular membrane. Because it is also important to cell division, it is not surprising that calcium is found in large concentrations in young undifferentiated meristematic cells that are dividing. This is why calcium deficiencies appear first in the growing areas of roots, shoots, and young leaves, and the result of the deficiency is that they become malformed.

Calcium is not mobile once assimilated into a cell. It is extremely abundant in the environment, and it comes from weathered minerals, such as limestone and chalk.

► **Magnesium** The key function of magnesium (Mg) is its role as the center of the chlorophyll molecule ($C_{55}H_{72}MgN_4O_5$). It is easy to see why this element is essential to plants: no magnesium, no photosynthesis. Magnesium is also an activator of enzymes involved in the production and use of ATP, so it plays a key role in respiration as well. Finally, magnesium is needed for the creation of DNA and RNA.

When plants develop a magnesium deficiency, chlorophyll can't be synthesized and photosynthesis ceases. Leaves start to lose their green color in between the leaf veins, which are nearest the remaining sources of sugars, whereas the veins remain green, a condition known as interveinal chlorosis. Older leaves show the signs of magnesium deficiency first, meaning that magnesium is mobile in plants.

Magnesium is a very abundant mineral on Earth, and it weathers out of many minerals. The most familiar form is dolomite ($CaMg[CO_3]_2$), a sedimentary carbonate rock. Much magnesium is dissolved in water.

► **Sulfur** Sulfur (S) is a component of two important amino acids, cysteine and methionine. Because of the orientation of a sulfur atom's bonds, proteins that include sulfur atoms take on particular shapes that influence their metabolic and structural roles. Methionine is found in structural components of a cell, and cysteine is important for metabolic activities. Cysteine is necessary to transfer electrons during photosynthesis and respiration, and it plays a role as a part of flavor molecules that can protect plants from grazing by insects and animals (including lots of humans, as haters of sulfur-filled broccoli cells know).

The yellowing that is a symptom of sulfur deficiency first appears in younger leaves

Sulfur is not mobile in plants, meaning that once it is synthesized into something, it essentially remains where it is. The yellowing that is a symptom of sulfur deficiency first appears in younger leaves. This is how you can distinguish a lack of sulfur from a nitrogen deficiency (although there are other causes of the same symptom).

There used to be quite a bit of sulfur in air pollution, and rains deposited this. This deposition would form sulfuric acid in bodies of water, hence the name acid rain. This deposited sulfur also provided a major source for plants, enough so that sulfur deficiencies were very rare. One of the results of stricter air pollution regulation has been a dramatic drop in sulfur in soils. Fortunately, sulfur is also released from organic material by microbial activity and it is weathered from gypsum rock, which can be mined.

► **Silicon** Many scientists do not consider silicon (Si) to be an essential element. I am including this seventeenth element here because there are some plants that can't live without it (those in the Equisetaceae family, known as horsetail or scouring rush).

Silicon improves drought tolerance by forming a waterproof layer in epidermal cells, thereby increasing the barrier and keeping water in. It also helps ward off insects. Many believe this is because silicon can make four bonds like carbon and thus can imitate carbon substances to block pathogens. Notice that silicon is grouped with the macronutrients. Those plants that do require silicon need it in larger quantities than trace amounts. Fortunately, over 90 percent of the Earth's crust is composed of silicate minerals.

Silicon is not mobile in a plant, so deficiencies show up on younger leaves. Again, however, silicon is not technically an essential element because most plants can grow and reproduce without it.

MICRONUTRIENTS

According to Von Liebig's Law of the Minimum, even if the least used nutrient is not present, it won't do any good to have all of the others. So, elements used in trace amounts are every bit as essential as the macronutrients. Without them, plants can't grow and reproduce.

► **Boron** Boron (B) is a component in cell walls, where it connects the carbohydrate chains. It is necessary for the formation of pollen tubes, and thus pollen doesn't form properly without this nutrient. Boron also plays a role in balancing the amounts of sugar versus starch in a plant, and it is involved in the translocation of both throughout the plant. Boron also helps transport potassium ions across the cell membrane, and thus helps to regulate the opening and closing of leaf stomata. This nutrient is usually taken up as boric acid ($B[OH]_3$) and uses three different transport mechanisms to cross cellular membranes: diffusion, facilitative transport, and active transport.

Boron is the only mineral nutrient that does not have to be an ion to move into plant cells

Boron is the only mineral nutrient that does not have to be an ion to move into plant cells. Once incorporated, boron is not mobile in plants because it readily bonds to sugar molecules. Symptoms of boron deficiency include damage at the growing tips and problems with flower and fruit formation. Borax ($Na_2B_4O_7$) is the ore of boron.

► **Chlorine** Chlorine (Cl) is a relative newcomer to the list of essential nutrients. Scientists have discovered that it is necessary for the operation of stomata and in maintaining the electrical balance of ions. Chlorine is also needed to break the bonds holding water molecules together during photosynthesis; it is the chlorine atoms that supply electrons for the process.

Chlorine in its ionic form, chloride (Cl^-), is the counter balance to the potassium ion (K^+) in operating stomata. If there isn't enough chloride, plant tips wilt and turn bronze. They often appear mottled with spots of chlorosis. Chlorine is not mobile in plants, so it is the newer leaves that show symptoms. Too much chlorine causes yellowing of the leaf margins, and this does occur in older leaves.

► **Copper** Copper (Cu) is a key component in enzymes used in oxidation and the reactions that occur during photosynthesis and respiration. Copper is also in enzymes that build amino acids into proteins, and it plays a key role in the synthesis of lignin.

Copper is immobile once assimilated. A lack of copper results in chlorosis (because it is needed in photosynthesis), the curling of leaves in some plants, and excessive branching in others.

► **Iron** Iron (Fe) atoms serve as carriers of electrons, and the electrons can be easily passed around between them. As such, iron is useful for respiration and oxidation-reduction reactions, where electrons are taken from one molecule and given to another. Iron is also used to make chlorophyll (although it is not part of the molecule) and helps it to function properly. Iron is critical to nitrogen fixation, as the conversion of atmospheric nitrogen and nitrate relies on iron. It is a component of important enzymes and proteins. In fact, iron is so important to plants that they have developed a system to release ions into the soil to lower the pH to prevent iron from becoming unavailable.

Plants that are low in iron show symptoms of chlorosis (yellowing), which is due to a loss of chlorophyll's green pigment. Iron is not a mobile nutrient in plants, so young leaves may become bleached, while older leaves and especially their veins continue to be green. Older leaves eventually lose color, especially at the margins, when there is a shortage of iron. Iron comes from mined minerals, and it is common in most soils.

► **Manganese** Manganese (Mn) is used to free oxygen during photosynthesis by accepting electrons from water. It also works with certain enzymes to break apart carbohydrates.

Manganese is not mobile in plants. Plants with low manganese show interveinal chlorosis, with yellow leaves and green veins.

► **Zinc** Zinc (Zn) is a component of lots of enzymes and is involved in donating or accepting electrons (oxidation and reduction). Zinc helps in the production of auxins, the main growth hormone in plants. It is needed to synthesize chlorophyll and carbohydrates, and it activates enzymes needed to make some proteins, particularly RNA and DNA. Zinc can also help plant cells withstand cold temperatures.

Zinc is a mobile element in plants. Its importance to growth is clear, as are the symptoms shown without it: slow and then no growth, lack of stem elongation, and yellowing.

► **Molybdenum** Molybdenum (Mo) is necessary for the synthesis of organic phosphorus compounds once phosphorus enters the plant. Also, while not directly related to cellular nutrition, molybdenum is needed in order for *Rhizobia* and *Frankia* bacteria to fix nitrogen in the root nodules of legumes.

Molybdenum is needed in order for *Rhizobia* and *Frankia* bacteria to fix nitrogen in the root nodules of legumes

Molybdenum is mobile, as evidenced by the chlorosis that develops in older leaves as this limiting nutrient is moved to new, growing tissues. One of the symptoms of too little molybdenum is a buildup of nitrate. This, in turn, causes some leaves to curl, a symptom called whiptail.

► **Nickel** There are many who still don't recognize nickel (Ni) as an essential nutrient because its functions are obscure. However, when nickel is absent, nitrogen in the form of urea accumulates in leaves. This results in leaf tip burns.

Nickel is an immobile element in plants. It is a mineral present in several different kinds of rocks.

► **Sodium** According to some scientists, sodium (Na) is not an essential element because it is not needed for all plants to grow. However, in C4 plants, which include sugarcane, maize, sorghum, and amaranth, sodium can help support osmotic activities in cells, drawing water into cells when there are not enough potassium ions to do the job. Other cations, such as those of the element rubidium, can do this as well. That is part of the reason why many do not consider sodium an essential element: there is another element that can do the same job. Sodium is mobile in those plants that require it.

GROUPING NUTRIENTS BASED ON BIOCHEMICAL FUNCTIONS

The essential nutrients have traditionally been grouped based on the amounts in which they are needed by plants. Because they are all necessary elements, however, it may make more sense to classify plant nutrients by their functions inside the plant.

By using this criterion, nitrogen, sulfur, phosphorus, and boron are grouped together because they are necessary for the formation of structural parts of plants. The first three are components of proteins, which are the building blocks of most structural elements. Boron is included because it provides strength to the polysaccharide connections inside cell walls. Nitrogen, sulfur, and phosphorus form another group whose electron bonding capabilities are important to their functions. Their bonds are involved in the storage of nutrients and creation of energy inside the cell.

The nutrients potassium, magnesium, calcium, manganese, zinc, iron, copper, and molybdenum all affect or activate enzymes in one form or another and serve as electron transporters. Phosphorus, chlorine, sodium, and potassium are grouped together because they affect the way membranes work. Then there are those nutrients needed for the electrical balance of ions inside and outside of cells. These are chlorine, potassium, calcium, and magnesium, all cell function regulators.

Using these classifications makes more sense when it comes to figuring out why deficiencies occur when mineral ions are missing from a plant's diet. Yes, there is duplication, but, hey, there are not many nutrients.

GROUPING NUTRIENTS BASED ON MOBILITY

Plant nutrients have varying degrees of mobility in the soil and once inside plants. Mobility, then, can also be used as a way to classify nutrients, and this classification system helps with our understanding of what they do.

Several factors influence the mobility of nutrients in the soil, including pH, moisture, clay, and organic matter, as well as their reactions with each other. In general, however, the most mobile nutrients in the soil are those that are highly soluble. These include nitrogen in the form of nitrate, sulfur, boron, and chlorine. Nitrogen in the form of ammonium, potassium, calcium, magnesium, molybdenum, and nickel are adsorbed by clay and organic matter and are less mobile. The really immobile nutrients react chemically and are tied up; these usually include phosphorus, copper, iron, manganese, and zinc, although these can become mobile when chelated.

The mobility of nutrients in plants is also variable and is influenced by the age of the plant, level of deficiency, and interactions with other nutrients, especially nitrogen. Generally, the most mobile nutrients are nitrogen, phosphorus, potassium, magnesium, and chlorine. When these nutrients are in short supply, the plant can transport them to areas of new growth, so symptoms of their deficiencies usually first appear in older parts of a plant. Iron, zinc, manganese, copper, nickel, sulfur, and molybdenum are somewhat immobile. In this case, symptoms of deficiency appear first on new growth, as they do with the very immobile nutrients, calcium and boron.

Nutrient mobility inside plants is one reason why foliar feeding is not always effective and is almost never an efficient way to fertilize a plant. Although it is possible for most nutrients to be absorbed through the foliage of plants, only the mobile ones can be moved to where they are to be used. In the case of macronutrients, it's impossible to supply what the plant really needs simply by spraying leaves.

In sum, nitrogen, phosphorus, potassium, magnesium, molybdenum, and nickel are usually mobile inside plants and deficiencies show up in lower, older leaves as these mobile nutrients are moved into newer leaves when they are in short supply. Calcium, boron, sulfur, iron, manganese, and copper are usually immobile once inside plants. Deficiencies of these immobile elements appear first in the youngest parts of the plant, the terminal buds and roots. Zinc deficiencies show up in the middle-aged leaves and deficiencies of chlorine, though mobile, appear in the top leaves. The mobility of boron is dependent on carrier molecules to which it binds. This situation points out the unreliability of determining nutrient deficiencies by just looking at a plant. All sorts of chemical interdependencies can mask the real deficiency, and only a test will tell what is going on.

Finally, it bears repeating that, except for boron, to pass through

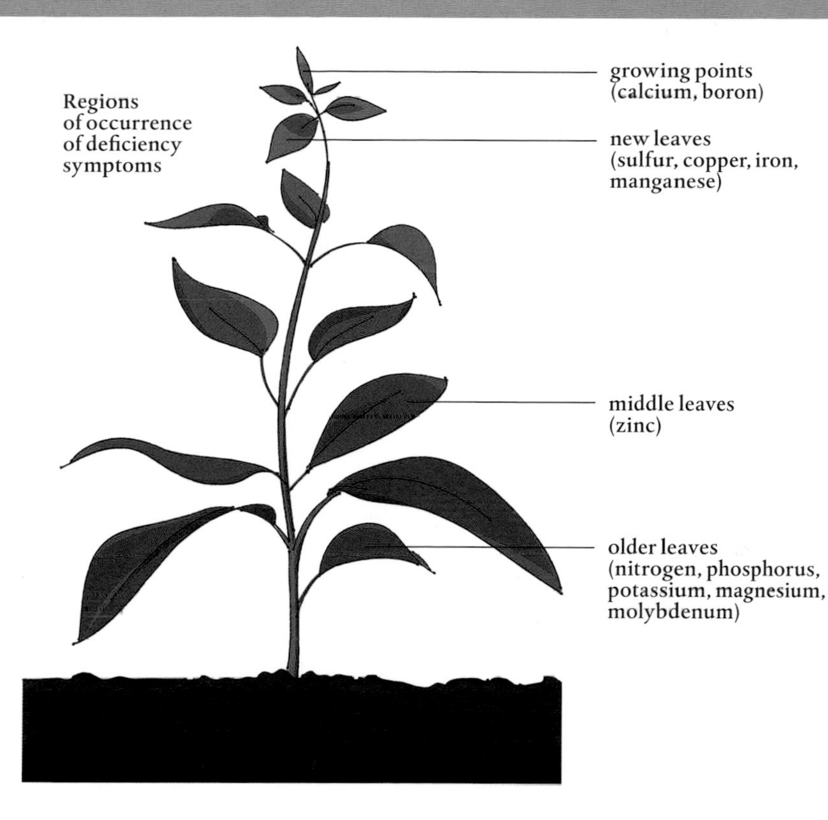

Deficiencies of essential elements show up in different locations of the plant due, in part, to their mobility. Symptoms in older leaves are usually associated with mobile nutrients, whereas those caused by less mobile nutrients usually occur in newer parts of the plant. One exception is chlorine, which is mobile, but symptoms usually appear in the upper leaves first. This is not an exact science, and other environmental factors have much to do with plant symptoms as well.

plant cell membranes, nutrients have to be in ionic form. If they are not, they cannot enter the plant. This is key to our story, obviously.

The mobility of these same nutrients in soil is of interest, too, as this mobility has a lot to do with availability and application. Of course, the soil characteristics come into play, but the ionic charges of nutrients are also important to how nutrients move in soil. The mobility of charged

THE IONIC FORMS OF ESSENTIAL NUTRIENTS REQUIRED FOR PLANT INTAKE

Macronutrients	Ionic forms	Micronutrients	Ionic forms
carbon (C)	—	chlorine (Cl)	Cl^-
hydrogen (H)	—	copper (Cu)	Cu^{+2}
oxygen (O)	—	iron (Fe)	Fe^{+2}, Fe^{+3}
nitrogen (N)	nitrate (NO_3^-), ammonium (NH_4^+)	manganese (Mn)	Mn^{+2}
phosphorus (P)	phosphates (HPO_4^{-2}, $H_2PO_4^-$)	zinc (Zn)	Zn^{+2}
potassium (K)	K^+	molybdenum (Mo)	MoO_4^{-2}
calcium (Ca)	Ca^{+2}	nickel (Ni)	Ni^{+2}
magnesium (Mg)	Mg^{+2}		
sulfur (S)	sulfate (SO_4^{-2})		

Note: Boron is taken up as boric acid (H_2BO_3), which is not an ion.

particles in soil depends on the charges (and their strengths) that exist on soil particles at cation and anion exchange sites. It also depends on the presence of water.

The ions in soil are located in two places. First, cations are attached to clay and humus particles that hold a negative charge on their surfaces. These negative charges attract the positively charged nutrient ions and adsorb them on their surfaces. Second, negatively charged nutrient anions usually remain in the water solution and do not attach to soil particles in most instances. One exception is phosphate anions, which often adhere to clay particles despite their negative charge. This is because of the nature of clay, which is composed of sheets of molecules. Some positively charged areas in these sheets become exposed on the clay surface, attracting phosphate. In many tropical soils, the clay is extremely weathered and sulfates, nitrates, and chlorine anions are also held in the soil. For most gardeners and farmers, however, it is phosphate binding to soil that is of most concern.

The surfaces of plant roots carry a charge because root cells produce hydrogen ions (H^+) to use in their quest for nutrients. These cations are exported from a root hair and traded for the nutrient cations adhering to the surfaces of the soil particles, a process known as cation exchange. The cation exchange capacity is the number of cations that can be exchanged by a soil, and this trait can be measured.

It's not all about cations, of course. Roots constantly produce hydroxyl ions (OH^-) as well as hydrogen ions and remove them from the cells in order to balance the pH in the cytoplasm. As they are released into the soil, these hydroxyl ions can participate in anion exchange. Most anions, however, are dissolved in soil water and are absorbed into the cell when they come into contact with root surfaces. They travel in the apoplastic pathway until they enter the cell and the symplastic pathway.

As you might expect, different powers of the charges on nutrient ions result in stronger or weaker attraction to soil and root surfaces and different solubility in water. This is why a calcium ion (Ca^{+2}), with two positive charges, binds more tightly than a potassium ion (K^+), which only has one. Note, too, that anions leach out of soil faster than cations because the former more readily dissolve in water.

Back to the point at hand: boron, chlorine, sulfur, and nitrogen in the form of NO_3^- are mobile in soil. On the other side of the spectrum, nitrogen in the form of NH_4^+, calcium, copper, iron, magnesium, manganese, molybdenum, nickel, phosphorus, potassium, and zinc are relatively immobile in soil. This is important when it comes to their application as fertilizers, obviously.

NONESSENTIAL ELEMENTS IN PLANTS

Clearly, plants contain other elements beyond the essential ones. How else would seaweed, which is nothing more than a plant, contain more than sixty elements? Arsenic, lead, gold, mercury, cobalt, uranium, and sodium are some of the nonessential elements taken up by plants through the plasmalemma via the same membrane transport proteins that take up the essential elements. These transporters are very specific as to what they will allow to pass through and have obviously

CATION EXCHANGE CAPACITY

A soil with higher clay and humus contents will have a higher cation exchange capacity. In general, the higher the cation exchange capacity, the higher the soil fertility.

Clay and organic matter have vast surface areas covered with negative electrical charges. Plant nutrients also have electrical charges. Those nutrients in the soil water solution that are cations (positively charged ions) are attracted to and attach to the negative sites on soil particles.

Root cells pump out hydrogen ions (H^+) to help transport nutrient molecules across the cellular membrane. These ions accumulate on the cell walls of root hairs, where they often come into contact with charged soil particles (either directly or in soil solution). The hydrogen ions exchange locations with the cations. Once the new cation attaches to the cell wall, it can travel the apoplastic pathway or enter the cell via the symplastic pathway.

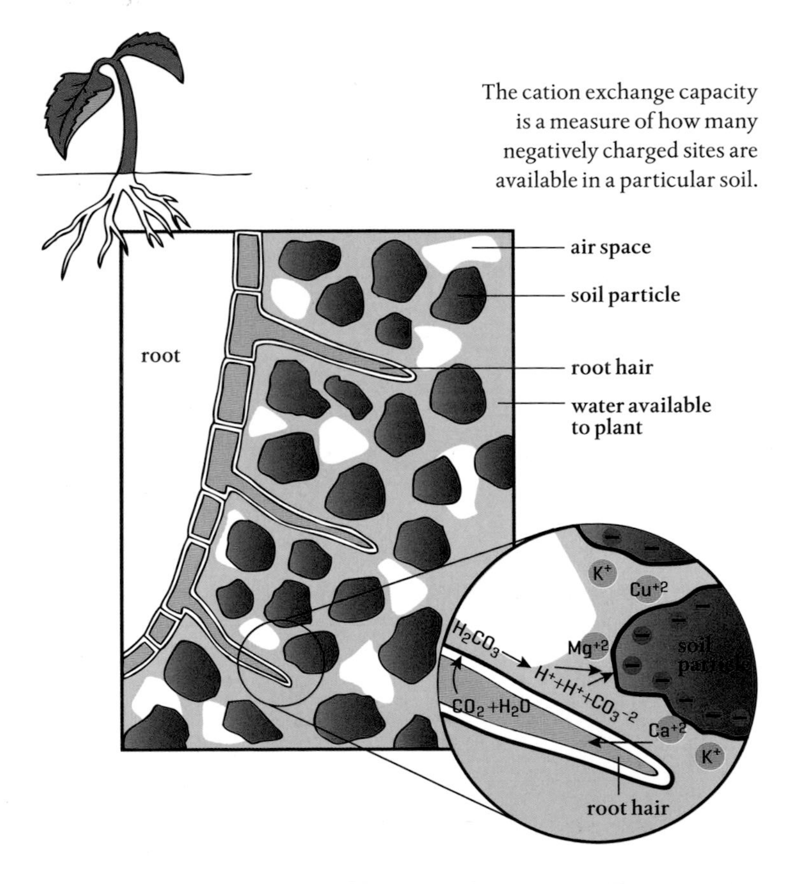

The cation exchange capacity is a measure of how many negatively charged sites are available in a particular soil.

evolved to center around letting in essential nutrients. Still, sometimes a nonessential element will slip through. If not used to make proteins or for some metabolic purposes, these nonessential elements are sequestered in vacuoles or converted to a insoluble form, such as a crystal.

Of course, the shape of the nonessential element, or the shape of the ion to which it is attached, has a lot to do with how it gets into a plant. Rice plants let in arsenite, which has a shape similar to that of silicic acid and is taken up by a silicon membrane transport protein. Arsenic (As) in the form of AsO_3^- apparently moves through the same membrane transport protein as phosphorus.

Some plants have developed the ability to take up quite a bit of nonessential nutrients that are toxic to other plants. For example, the fern *Pteris vittata* takes up large quantities of arsenic. This can be in the form of hyperaccumulation, wherein a specific metal makes up 1 percent or more of the dry mass of the plant. Or it can take place via photovolatilization, in which the nonessential element is turned into a gas and released from the plant via transpiration. In phytoremediation, plants (and sometimes the microbes they attract) are used to detoxify soil contaminated with materials such as heavy metals, pesticides, and crude oil.

In general, mycorrhizal fungi bring metal nutrients such as phosphorus, copper, and nickel to plants. Some can also bring lead, which is not essential to plants. Lead can be accumulated in plants that support such fungi. Mycorrhizal fungi can also sequester nonessential nutrients, accumulating them so they won't harm their symbiotic partner.

Why would a plant take up a nonessential element? After all, the synthesis and maintenance of the transporter proteins that regulate what can cross the cell membrane take up valuable energy resources. Still, up to two-thirds of the ninety naturally occurring elements can be found in plants, and in many cases these nonessential nutrients stimulate chlorophyll content. Apparently, something can be a beneficial element and yet not an essential one. Or it could be that some of these are essential, but in such small quantities as to be impossible to measure. Undoubtedly, with the increasing ability to measure smaller and smaller amounts of a substance, some of these elements will be declared essential. Fortunately, they are needed in such small quantities that we gardeners would have no need to supply them.

Being nonessential is not the same thing as not being useful. Tea plants apparently have higher beneficial antioxidant properties when grown in the presence of aluminum. The element aluminum can act like a coenzyme, helping a catalytic process along. Low levels of cobalt have been found to aid legume growth. Silicon helps cucumbers and roses grow. Selenium, in trace amounts, helps lettuce, potatoes, and rye. Titanium helps maize and other food crops by stimulating enzymatic activity, which increases the rate of photosynthesis and the uptake of nutrients.

Plants have actually developed some pretty neat systems to minimize the uptake of nonessential elements. Metals, for example, are complexed in the soil by a plant's exudates before they can be taken up. Organic acids can chelate metals (gather them up) and make them unavailable. In addition, some plants partner with arbuscular fungi that secrete glomalin, a compound that ties up metals in the soil.

SUMMING UP

Only seventeen elements are needed by plants to grow and reproduce. Now that we have identified and briefly discussed them, the difficult work is done and the exciting part begins.

KEY POINTS

- **Essential nutrients** are those elements required by all plants to grow and reproduce.
- **Macronutrients** are those required in large quantities by plants, whereas micronutrients are needed in only small or trace amounts.
- **Water and carbon dioxide** are the sources of a plant's hydrogen, oxygen, and carbon, which account for 96 percent of the mass of a plant.
- **Nitrogen is essential** because it is part of proteins, such as enzymes; nucleotides, which make up DNA and RNA molecules; and chlorophyll, which is required for photosynthesis.

- **Phosphorus** is part of adenosine triphosphate, DNA, and RNA molecules. It is also included in the phospholipid membrane that controls what enters a cell.
- **Potassium ions** play a key role in the movement of water into and out of cells.
- **Calcium** is a key structural component of cell walls and many enzymes, and it is important as a signaling molecule.
- **Magnesium** forms the center of the chlorophyll molecule and is thus essential for photosynthesis.
- **Sulfur is a component** of the amino acids cysteine and methionine, which are found in structural cell components and play a role in important metabolic activities.
- **Silicon** improves drought tolerance by forming a waterproof layer in epidermal cells, and it helps ward off insects.
- **Boron** is in cell walls, necessary for the formation of pollen tubes and involved in moving starch and sugar throughout a plant. It is the only nutrient that can move across cellular membranes as a non-charged molecule.
- **Chlorine** is needed for the operation of stomata and to break apart water molecules during photosynthesis.
- **Copper** is a key component in enzymes used in oxidation, photosynthesis, respiration, and the construction of proteins.
- **Iron atoms** serve as carriers of electrons, so iron is useful in respiration and oxidation-reduction reactions. Iron is used to make chlorophyll and is a component of important enzymes and proteins.
- **Manganese** is used to free oxygen during

photosynthesis and it works with enzymes to break apart carbohydrates.

- **Zinc** is part of many enzymes. It aids in the production of growth hormones, chlorophyll, and carbohydrates, and it activates enzymes needed to make RNA and DNA.
- **Molybdenum** is necessary for the synthesis of organic phosphorus compounds.
- **Nickel** is needed for the removal of urea, a nitrogen-based waste, from leaves.
- **In the soil,** nitrogen (as ammonium), potassium, calcium, and magnesium are more mobile than copper, iron, manganese, nickel, and zinc.
- **Deficiencies** of nutrients that are mobile within plants tend to show symptoms in older leaves, whereas newer leaves and growing tips show signs of deficiencies in less mobile or immobile nutrients.

FIVE

Water Movement through Plants

WHAT'S IN THIS CHAPTER

- **Roots,** and specifically root hairs, intercept water in the soil.
- **In the process** of mass flow, hydrogen-bonded water molecules are pulled like a chain out of the soil and into the plant.
- **Root pressure** exists when there is more water outside the root than inside it. This difference causes water to flow by osmosis into the root.
- **Once inside** the plant, water and nutrient ions are loaded into the xylem system and carried up to the leaves.
- **Water also transports** sugars made in the leaves, as well as the thousands of various proteins, enzymes, and hormones produced in plant cells, throughout the plant in the phloem system.

UP TO 90 PERCENT of a plant's weight is water in its cells and cell walls. We gardeners spend a lot of time and money to ensure that our plants get enough water. In fact, if nature doesn't supply the right amount of water, we must.

Water is a pretty special molecule. A water molecule is polar—that is, it has opposite electrical charges at either end. Its unique properties make water the actor with the leading role in our play. Of key import, water is the universal solvent. More kinds of molecules dissolve when put into water than any other liquid. It has myriads of hydrogen bonds that, while individually weak, strongly hold water molecules together. Also, water has the ability to cause other compounds to react with it and change in some way. For all of these reasons, water is the medium in which plant nutrients move.

The journey water takes from the soil into plant roots has everything to do with how plants feed themselves. Water transports nutrients, in ionic form, into plants. Throughout their time inside the plant, it is water that distributes nutrients and the products into which they are synthesized to where they are needed.

HOW WATER GETS TO ROOTS

We've jumped the gun just a bit. We first need to understand how water gets to roots in the first place. There are two systems at play. Some roots grow into areas of soil that contain water and thus come into direct contact with it by interception. There is usually a microscopic film of water around soil particles. In any case, roots and specifically root hairs intercept water.

In most instances, however, mass flow is responsible for water getting to a root. Mass flow is what happens when you put a dry sponge on a puddle of water. The puddle flows toward the sponge and eventually up into it. This is exactly what happens in the soil, only the root is the sponge. Mass flow occurs because of the hydrogen bonds between water molecules. Just as water molecules pull each other through the xylem during transpiration, water molecules are pulled by hydrogen-bonded neighbors out of the soil and into the plant. This causes those water molecules behind them in the soil to be pulled closer and toward the plant. The adhesive and cohesive powers of water molecules are at work, and mass flow is created.

THE MOVEMENT OF WATER INTO A ROOT

The next question is what causes water molecules to move into a root. Again, there are two answers. First, transpiration takes over. About 90 percent of a plant's water loss happens as a result of evaporation through leaf stomata. This suggests that as much as 90 percent of a plant's water comes into the roots as a result of transpiration, pulled molecule by hydrogen-bonded molecule into, up, and out of the plant at the leaf surface. The molecules evaporating from the leaf surface pull up water from below. This, in turn, causes a pressure deficit in the roots that pulls in more water from the soil.

Lots of water moves through a plant due to transpiration. A mature maize plant, for example, transpires about 4 gallons (15 liters) per week. That means 1 acre (0.4 hectare) of maize transpires 350,000 gallons (1.3 million liters) during its 100-day growing season. The birch trees in my yard in Alaska each transpire 200 to 1000 gallons (760 to 3800 liters) per week. To know all of this is regulated by tiny stomata cells embedded in leaf epidermis, which, in turn, are regulated by potassium and chlorine makes it even that more wonderful, in the truest sense of that word.

The second method by which water enters plants is via absorption as a result of root pressure. Root pressure is a common phenomenon that occurs when there is more water outside the root than inside it. This difference causes water to flow in the direction that will create a state of equilibrium—that is, into the root. This pressure is aided by higher concentrations of dissolved ions in the root cells than in the water outside them, causing water to flow into the root to dilute them (osmosis).

A good example of the root pressure phenomenon is imbibition, the swelling that occurs in a seed as a result of water absorption. Water flows in to dilute the concentrates inside the seed. Imbibition can create tremendous force inside a seed, 1000 times stronger than atmospheric pressures and enough so that a single seed can crack open a rock crevice as it swells.

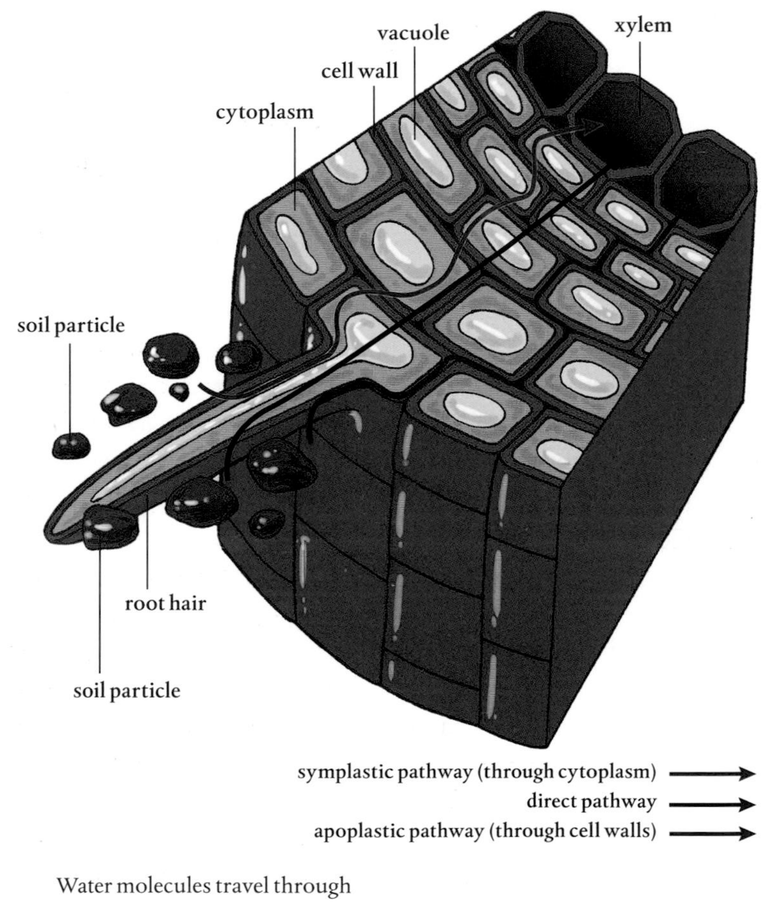

Water molecules travel through plants along three pathways.

TRANSPORT IN THE APOPLASTIC AND SYMPLASTIC PATHWAYS TO THE XYLEM

If you've been keeping notes, you know that water molecules can travel in plants three different ways: the apoplastic, symplastic, and intercellular pathways. This triple redundancy reflects water's importance to plants, and it allows a plant to maintain the amount of water that it requires. Although there are three separate pathways for water once it gets inside the cell wall of a root hair, all three are put to use at the same time.

Let's start with the apoplastic pathway created by the porous cell walls and intercellular spaces around plant cells. Together, these create a continuous space, not only around each cell, but also from one cell to the next, with few exceptions, throughout the plant. This pathway is a porous space between the cellulose fibers that allows water to travel into a plant without actually entering any cells. Cell walls account for up to 20 percent of a plant, meaning that they provide a lot of space inside a plant that can hold a lot of water. The apoplastic pathway allows water and the nutrients it carries to get far into the root without having to deal with the much more restrictive plasmalemma barriers that would regulate it, and particularly its solutes, as they enter cells.

The apoplastic pathway hits a dead end, of course, when it comes to the waterproof Casparian strip, the single layer of wax-clogged cells at the endodermis of a root. At this point, water molecules either finally enter the symplastic pathway by crossing the plasmalemma or move back out of the root via the cell walls in reverse. This way, the plasmalemma regulates everything that goes into the cell.

Of course, water molecules that take the apoplastic pathway have the option of diffusing into a cell through the plasmalemma—that is, entering the symplastic pathway—anywhere along their route. Water can pass directly through cellular membranes by osmosis, with its small molecules slipping in between the lipid tails of the plasma membrane. Water can also cross the cell membrane via the special water transport proteins, aquaporins, embedded in the membrane.

In the symplastic pathway, water molecules move in a space defined by the inner side of the plasmalemma and the outside the tonoplast membrane that surrounds the vacuole. This is the cytosol. The symplastic pathway extends from cell to cell throughout the plant because plasmodesmata tunnels open every cell to its neighbors.

Although all living plant cells have plasmodesmata, some have more than others. This variation depends on the function of the particular cell. One involved with transporting water, say, right up next to a root hair, might have 10 to 15 pores per square micron. In the case of a single young barley root cell, the number of plasmodesmata is about 20,000. Once these cells mature, they have fewer plasmodesmata that can function. When there is a need to tightly regulate ion concentrations in the cytosol, the numbers of functional plasmodesmata decrease for more precise control. Some plasmodesmata are simple linear tunnels, others are H-shaped, and some are branched (though usually only in mature plants). They are in the range of 20 to 60 nanometers wide. Obviously, there is some regulation afforded by the different sizes and internal shapes.

A small tube known as the desmotubule is found inside each plasmodesma and may provide support for it and be involved with the size of molecules that are allowed to pass. This tube is connected to both cells' endoplasmic reticulums, where the synthesis of proteins and lipids is completed. Studies suggest that the desmotubule may allow ions another mode of transport through cells (other than transport proteins) when conditions merit. It also may allow the transport of large protein molecules between two cells.

Water is such a unique molecule that it can pass through all plant membranes. Because it doesn't need a plasmodesma tunnel or an aquaporin, it can go through them all in a straight line, along the intercellular pathway. Again, this is the direct result of the wonderful properties of water molecules and their ability to slip in between the phospholipid molecules forming a cell membrane.

LOADING AND UNLOADING THE XYLEM

Once in the plant root, water is pulled toward the xylem tissue. There, water molecules diffuse across the membranes of the living part of the xylem, the xylem parenchyma cells. These are situated next to the xylem tracheids and vessel elements, which are dead xylem cells. These cells have no cytoplasm, no cytosol, and, most importantly, no plasmalemma or plasmodesmata. Because no symplastic connection exists between tracheids and vessel elements, there is very little restrictions to flow.

In order for water and all the nutrient ions dissolved in it to get into xylem tubes, they have to be loaded into either a xylem tracheid or vessel element. To help, the membrane of the xylem parenchyma cells are full of aquaporins, ion carriers, ion channels, and ion pumps to move water and its nutrient solutes into the xylem system.

Let's follow some water molecules into the xylem. The molecules are pulled up from the soil and enter the plant through a root hair, and then move through the epidermis and into the cortex via the three different paths. At the endodermis, water molecules hit the Casparian strip and are forced to move into the symplastic pathway. Once near the xylem system, however, water molecules have to diffuse out of the symplastic and into the apoplastic pathway, which is the only connection with the dead xylem vessels and tracheids. The molecules then move up the xylem system, pulled by transpiration toward the stomata.

The xylem is an amazing plumbing system

The xylem is an amazing plumbing system, even if its flow is unidirectional. The xylem tissue runs from the root system through the stem and into the veins of the leaves. All along the way, water and nutrients can unload and diffuse or be otherwise transported into cells, where they are used to create energy or to build cell parts.

As some plants grow, the oldest xylem no longer functions. New xylem cells at the pericycle replace it. Layers of old xylem make the rings in trees. Interestingly, the size of xylem cells varies depending on environmental conditions. When it is wet in the spring, large cells are produced. As rains dissipate and as temperatures go up, thus increasing transpiration, thinner tubes are developed. This is why tree rings vary in thickness and are helpful to climatologists, serving as a record of precipitation and temperature in the past.

WATER TRANSPORTS ORGANIC COMPOUNDS IN THE PHLOEM

Plants are autotrophic, meaning that they make everything needed to sustain themselves. This starts with cells that can photosynthesize and incorporate carbon. These cells utilize the phloem to deliver the carbon-containing sugars they make to cells that cannot make these substances, along with other molecules picked up along the way.

Phloem sieve tubes are not typical cells. They lack nuclei, vacuoles, and plastids and only have a limited number of mitochondria. This allows sap to flow easier because there are fewer obstructions. Each vessel does have a plasmalemma, which covers even the sieve plates, the perforated structures that separate tubes. Each vessel has a companion cell with all the organelles of a normal cell. Lots of plasmodesmata connect the sieve tubes to their companion cells, and many of the large molecules that are carried to the sieve tubes pass through these.

The companion cell and the sieve tube it accompanies are both produced by the same mother cell. Instead of the twin cells produced during normal division, two very different offspring are created. Because it contains a nucleus, the companion cell controls the activities of the sieve tube. Companion cells have lots of mitochondria, which are needed to produce energy for both cells. Finally, the sieve tubes do not have direct connections to other cells; they only share plasmodesmata with companion cells.

There are also storage cells associated with the phloem. This storage

A sieve tube and its companion cell

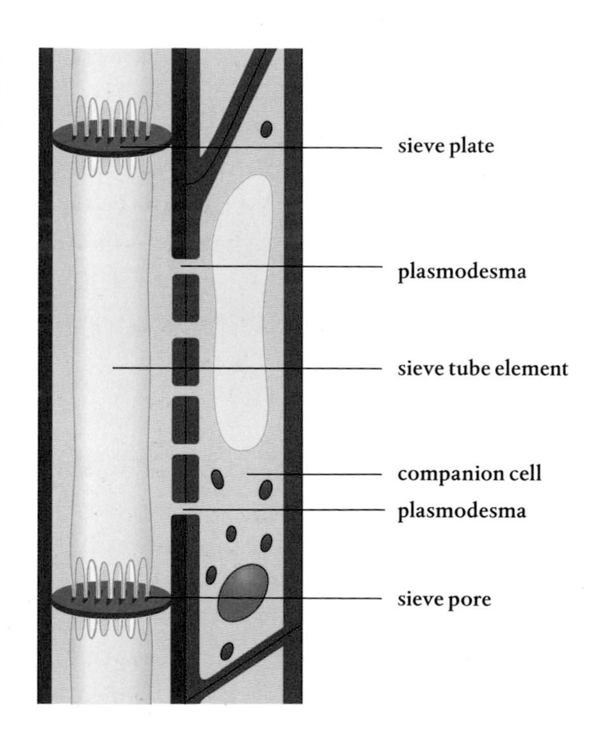

system extends from the veins of leaves all the way to the roots, where lots of energy is needed for growth. Flow is bidirectional. In the spring, for example, starches stored in the roots during autumn and winter are needed to get things going, and the flow in the phloem is upward. As autumn sets in and winter approaches, perennials, trees, and shrubs store many more starches in the roots than during the main growing season, and the flow is downward.

Like the xylem cells, phloem cells are connected end to end to form tubes. The ends of individual phloem cells, sieve plates, have very fine openings. When a plant is damaged, these plates become clogged with proteins, blocking off the damaged portion of a plant. This action prevents the loss of valuable sap. These plates clog only when there is damage.

Getting sugars into the phloem requires energy. These molecules are actively transported into the phloem in a process known as phloem loading. Some plants are apoplastic loaders, and others are symplastic loaders. You know about these two pathways and can correctly guess the routes. Both pathways lead to the phloem companion cells. At that point, energy is employed to move sugars and other molecules across the plasmalemma into the companion cell.

The symplastic loaders pass sugars and amino acids into the phloem companion cell though one or more of the many plasmodesmata in the plasmalemma. (See how this stuff just slides off your tongue by now.) Lots of transport proteins specific for either sucrose or proteins are embedded in the companion cell membrane to facilitate this transport.

The apoplastic loaders pass sugars and proteins into the apoplastic pathway. Hydrogen ions pumped out of the cell membrane and into the cell wall are used to then allow molecules into the companion cells.

The path taken into the phloem depends on lots of environmental factors. In general, the apoplastic pathway is widely used in temperate climates, whereas plants in the tropics rely more on the symplastic route. Sucrose travels along the apoplastic pathway and other sugars the symplastic. Most trees load the phloem via the symplastic pathway, whereas most herbaceous plants go with the apoplastic one.

In some plants the phloem dies every year, whereas in others the cells live for more than a year. Phloem systems in palm trees often survive for more than 50 years. When they do die, meristematic cells at the root apical meristem replace them.

The xylem and the phloem are related in function at times. Sugars move into the phloem via active transport from a source, resulting in a higher concentration inside the sieve tube. This causes some of the water molecules to move from the adjacent xylem into the sieve tube to dilute the sugar concentration. This movement of water into the restricted sieve tube creates turgor, which causes the water and solutes to flow through the system.

Somewhere along the phloem tube is a sink, an area where there is a lower concentration of sugars because they have been removed and utilized by nearby cells. This lower concentration of sugars in the solution causes the high-concentration phloem sap, which is 80 percent sugar and already under pressure, to move to the sink cells.

Once sap is at the sink, active transport mechanisms unload the sugars from the phloem into cells, where they are either used or moved

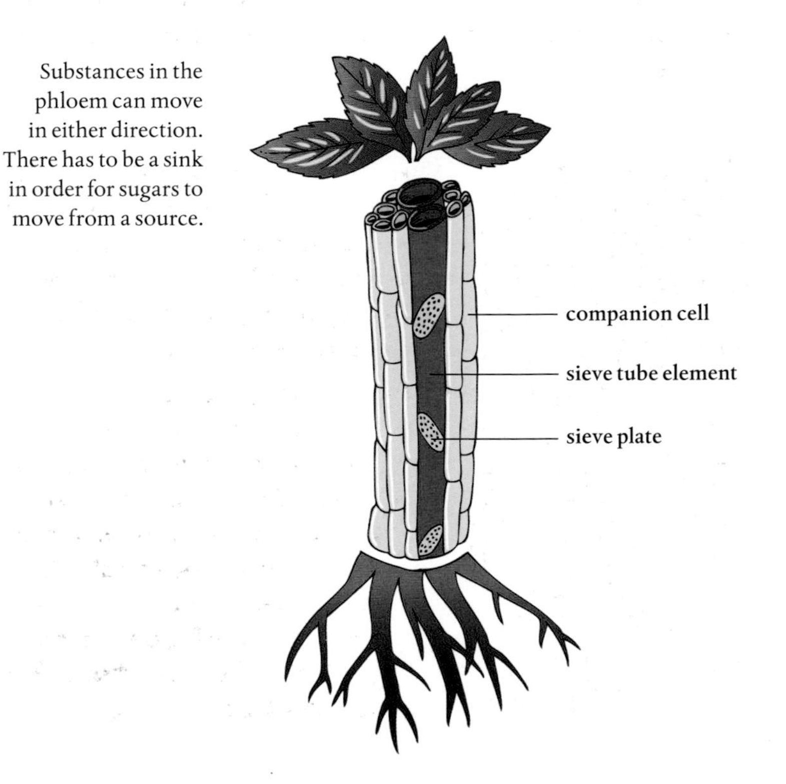

Substances in the phloem can move in either direction. There has to be a sink in order for sugars to move from a source.

to nearby cells for use. Some sugars move into the storage vacuoles of receiving cells, in each case crossing the tonoplast via specialized transport proteins. This has the effect of lowering the concentration of the sugars in the cytosol of the sink cells, which causes more sugars to flow into the cells.

Meanwhile, the water carrying concentrates to a sink loses sugars and flows out of the phloem and back to the xylem as a result of transpiration. Once free of the phloem, the transport of materials can continue from cell to cell, rather than through cell walls. Sink cells near the phloem usually have plasmodesmata that allow larger sugar and organic molecules to move from cell to cell.

SUMMING UP

Water is the source of life, and plants have evolved elegant systems to take advantage of its unusual properties. We gardeners provide our plants with plenty of water not only because it bathes the cells and keeps the soil food web alive, but because it carries with it the essential nutrients.

- **Water in the soil** gets to a plant by being intercepted by growing roots and because of mass flow, in which water molecules are pulled like a chain by hydrogen-bonded neighbors through the soil.
- **During transpiration,** water molecules evaporating from the leaf surface pull up water molecules from the xylem. This occurs as a result of cohesion of the water molecules to each other as well as adhesion of the molecules to the surfaces of the xylem, which causes a pressure deficit in the roots that pulls in more water from the soil.

- **Root pressure** occurs when there is more water outside the root than inside it. This difference causes water to flow into the root to equalize the concentrations.
- **The apoplastic pathway** is a porous space between the cellulose fibers of cell walls that allows water to travel into a plant without actually entering any cells.
- **The waterproof Casparian strip** in the endodermis of a root causes water to leave the apoplastic pathway and to pass through cell membranes or leave the root.
- **The symplastic pathway** is guarded by the plasmalemma, which regulates what can enter a cell.
- **Water and the nutrients** dissolved in it are loaded into the xylem tracheids and vessel elements by the xylem parenchyma cells, whose membranes have numerous aquaporins and ion carriers, channels, and pumps.
- **Phloem sieve tubes** transport the sap (water and dissolved substances produced by the plant) between the leaves, which produce sugars, and roots, which serve as storage organs.

SIX

Nutrient Movement through Plants

WHAT'S IN THIS CHAPTER

- **The electrical charges** on soil particles play a role in the movement of nutrient ions into a plant.
- **Microbes** are also essential in this process, and roots exude carbon-rich materials into the soil to attract these microbial partners.
- **Some of these microbes** live within plants, in root nodules and root hairs.
- **Others break down** nutrient-containing compounds in the soil, changing them into the ionic forms necessary for their uptake by plant roots.
- **Once at the root,** nutrient ions must diffuse or be actively transported across cell membranes so they can be used to make the compounds needed for growth and maintenance.

BY NOW YOU should have picked up on a theme: the essential plant nutrients move into the plant in conjunction with water. It shouldn't come as any surprise, then, that much of what I've described about water movement in the soil through root hairs and ultimately into the vascular tissues applies to nutrient movement as well.

There is more to it, of course, than just hitchhiking a ride with water, going with the flow, as it were. The essential mineral nutrients are ions with various charges. They also have different shapes and, as many gardeners know, are available to plants in different quantities depending on the characteristics of the soil.

GETTING NUTRIENTS TO THE ROOTS

Before broaching the process of how nutrients enter the root, it's a good idea to review how nutrients get to the roots. Yes, water plays a big role, but it does not act alone.

► **Interception, Mass Flow, and Diffusion** For starters, as roots grow they come into direct physical contact with nutrient ions—not just those in water, but also those that are loosely attached to the surfaces of clay particles and bits of organic matter. This is interception. For the most part, clay particles and organic matter in the soil carry a slight negative charge that attracts and loosely holds nutrient cations. Roots bear a negative charge, too, because hydrogen ions (H^+) are pumped out of cell membranes in order to set up active transport. Not all of these hydrogen ions make it back into the root. They are sometimes exchanged for the loosely attached cations in the soil. Nitrogen, potassium, calcium, magnesium, and copper all form positively charged ions that tend to adhere to soil particles.

Root interception accounts for an awfully small amount of nutrient uptake, however. It's estimated to cover only about 1 percent of the plant's needs. In addition, a cation depletion zone quickly forms around a root hair after cation exchange takes place. There is nothing else to exchange. Plant roots must continue to grow and new root hairs must develop to encounter new soil particles. This problem is lessened by association with mycorrhizal fungi, as described in the next section.

Negatively charged nutrient anions are generally not adsorbed on soil surfaces, and they are much more soluble in water as compared to cations. These anions, along with some nutrient cations that eventually dissolve in water, are pulled to the plant during mass flow, as discussed in the previous chapter. Hitchhiking with water drawn to the roots by mass flow is how nitrogen in the form of nitrate (NO_3^-), sulfur, boron, chlorine, and manganese get to the roots. (Note that nitrogen is available as both a positive and a negative ion.)

Mass flow is a double-edged sword, however. It is a good thing if plants can capture all the water with its ions. It is a bad thing when there are more nutrients in the water than the plants in the system can utilize and those nutrients end up polluting lakes, streams, and water tables.

Finally, diffusion plays a role in getting nutrients to plants. This occurs when nutrient ions move toward roots because the area has a lower concentration of these molecules. Again, these ions are suspended in water or very loosely attached to soil particles. If concentrations get low enough nearby, the ions will move out of an area of higher concentration into the area of lower concentration. Tiny amounts of nutrients attached to soil particles go into solution in soil water when the concentration of the molecules is low enough in the water. These are then swept to the plant roots by mass flow. Diffusion takes place right next to the root as a direct result of ions being absorbed into the root. The solution in the soil then has a lower concentration of ions and replacements move in. This simple diffusion is the way zinc, phosphorus, and potassium, in particular, move to the root surface.

In sum, nutrients get to plants in three ways. Plants can intercept some of their nutrients in the soil. The intercepted nutrients are calcium, magnesium, zinc, and manganese. Mass flow moves anions dissolved in water, and this is how nitrogen in the form of nitrate moves to roots. In addition, calcium, chlorine, magnesium, sulfur, copper, manganese, molybdenum, and nickel are transported to plant roots by mass flow. Phosphorus, potassium, zinc, and iron arrive at roots as a result of diffusion, which is a slower process than mass flow. For the most part, these move to the root via diffusion when neighboring concentrations drop due to root absorption. Boron can travel to plant roots via both mass flow and diffusion.

► **Microbial Partners** Plants can also enlist microbial partners to move nutrients to their roots. To initiate these partnerships, roots exude lipids and carbon-based molecules into the soil. These materials, which are produced by organelles in the roots cells, cross through the double-layered cellular membranes and move through the cell walls and out into the soil immediately surrounding the roots. Here in the rhizosphere, the increase of carbon, together with the carbon in mucilage and dead root tip cells, attracts bacteria and fungi and helps sustain populations near the root surfaces. For example, *Rhizobia* and *Frankia* are nitrogen-fixing bacteria that form relationships with legumes.

Mycorrhizal fungi form a symbiotic relationship with plants and get

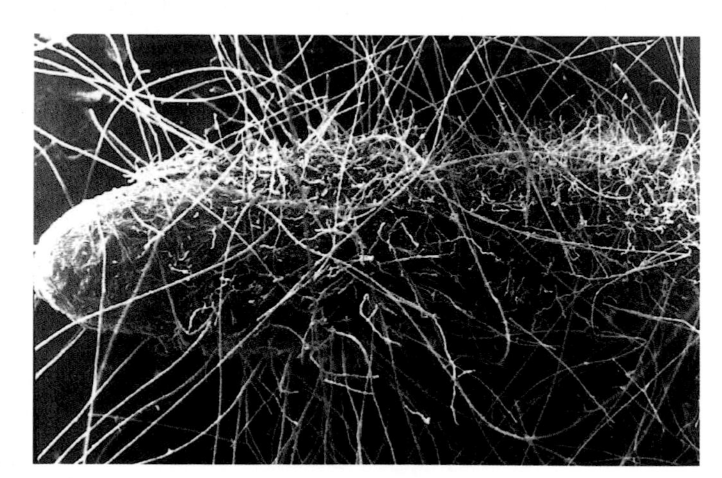

The long hyphae of mycorrhizal fungi help plants obtain nutrients and water. In return, the fungus receives exudates from the plant.

carbon directly from roots, ensuring a supply by providing the plant with nutrients. These fungi literally go out from the plant root hairs and obtain nutrients the roots cannot, both because of the inability of the larger root to fit into tiny pore spaces and the inability of the root to grow the length required.

These all-important root-fungi partnerships are known as mycorrhizae. (The fungi are mycorrhizal fungi; *mycorrhiza* refers to the relationship.) Somewhere between 90 and 96 percent of all plants enter into mycorrhizae. These are not trivial associations. Fungi produce powerful acids and enzymes that break down organic material, freeing tied-up nutrients, many of which are then moved to the host plant root. The plant, in turn, expends valuable energy and uses limited resources to make and deliver exudates to support the association.

Mycorrhizal fungi are best known for freeing phosphorus, which is so tightly bound to soil particles that little of it is readily available to plants. Although phosphorus is probably the nutrient most moved by them, nitrogen, copper, and zinc are also brought to the root by mycorrhizal fungi. (These guys will do almost anything in return for carbon.)

As more and more is learned about the role of mycorrhizae, it is becoming even clearer that the fungal partner plays a very important part when it comes to plant nutrients. For example, the number of nitrogen-fixing nodules in legumes increases in the presence of mycorrhizal fungi, and nitrogen released into the soil is retrieved and brought to the plant by the fungi.

The problem with plants enlisting biological help to obtain essential nutrients is that the process is not without its costs. The plant must expend energy and use limited resources to produce exudates. While most mycorrhizal fungi seem to form associations with more than one kind of plant, bacterial associations are quite specific to plant species as a result of the specific exudates produced by the plant to attract and support the necessary microbes. This dedication of DNA and RNA to the task of producing the necessary and specific plant exudates must be expensive. Perhaps this is why the specificity evolved. If you are going to associate, associate with the right organism.

Of course, in a gardener's plant-centered world, all the creatures of the soil food web are involved in bringing nutrients to plants. The arthropods and other invertebrates of the soil food web aid in the decay of organic material, releasing some and assimilating other nutrients in their bodies until they, in turn, are decayed. Eventually the nutrients from the soil food web organisms as well as decayed plant matter enter the bodies of bacteria, Archaea, and fungi. When these are eaten, the resultant wastes, in the form of inorganic ions, become available to plants via interception, mass flow, and diffusion. Step back and admire this system!

GETTING THROUGH THE MEMBRANE BARRIER

How do these nutrient ions get into the plant once they are at the plant root? It wasn't that long ago that scientists theorized plant roots engaged in eating soil organic matter to obtain nutrients. This was what Jethro Tull, the father of rototilling, and a whole generation of farmer-scientists believed until proven wrong by Justus Von Liebig in the mid-1800s, when he showed that plants take up nutrients in inorganic forms.

You, however, already know the answer. Once in appropriate ionic form, nutrients move into a root cell wall, usually at a root hair, and ultimately cross the plasma membrane to enter the cell's jelly-like cytosol. The plant creates the chemistry that allows it to obtain the nutrients it needs. The key to plant eating is the root cell membrane.

► **Passive Transport** The size of a molecule has a lot to do with how easily it can cross a membrane, if at all. It has to be small to diffuse directly through the membrane, or it has to fit into the transmembrane protein tunnels. Similarly, how an ion reacts to the lipid layer inside the cell membrane has a lot to do with how, or even if, the ion gets into a cell. If an ion dissolves in water, then it doesn't dissolve in lipids. This means it can't get through the phospholipid membrane without the aid of transport proteins.

Some plant nutrients move across all plant membranes with ease as a result of simple diffusion. This is passive transport, because no energy needs to be supplied for diffusion to occur. These molecules include oxygen, carbon dioxide, and water (although with water the process is osmosis). The concentration of any of these can vary on either side of a membrane. The natural force that seeks to equilibrate the concentration of molecules will result in flow toward the region of lower concentration.

In most instances, there is a concentration gradient between water outside a root and inside a root cell. The more molecules on one side, the higher the gradient and greater the tendency to move to the other side. Water outside the cell moves across the membrane into the cell cytosol, which is filled with water but also lots of solutes and organelles. In other words, the concentration of water is lower inside the cell. Keep in mind that diffusion and osmosis are forms of passive transport. No energy is added. Consider that it only takes a difference between the two sides of a few molecules, just one technically, to create an energy gradient.

► **Active Transport** Plants are constantly accumulating nutrients for growth and maintenance. Normally, concentrations of nutrients are greater inside a root cell than outside in the soil. For example, the concentration of phosphorus ions inside a root cell is thousands of times higher than it is in the soil because plants accumulate them. Now the energy gradient moves in the opposite direction, toward the soil. Fortunately, the plasmalemma can prevent or greatly slow diffusion and osmosis in the reverse direction.

The plant expends energy to actively transport ions from the soil across its root cell membranes against the energy gradient

Because the plant needs to continue to accumulate needed nutrients despite their being in higher concentrations inside the cell, the plant expends energy to actively transport ions from the soil across its root cell membranes against the energy gradient. This costs the plant energy. It's just like bicycling uphill versus coasting downhill.

Let's digress. Nutrient molecules are charged, however, which means they are attracted to the opposite charge. When an ion moves in or out of a cell or vacuole, if it isn't matched by its opposite charge, an electrical voltage is created across the membrane. This means that plant nutrient ions move down an electro-chemical gradient, a combination of the potentials created by charges and molecular concentrations. To complicate things, there is another gradient. This is the pH gradient created by differences in the concentration of hydrogen ions (H^+) versus the concentration of hydroxyl ions (OH^-).

In any case, there is good evidence of active transport taking place in root cells. When the powerhouses that produce ATP are stopped artificially, some nutrients do not enter the plant because there are no hydrogen ions being produced and pumped out of the cell. Clearly, then, energy is being used to transport these nutrient ions across the living membrane.

Scientists have determined how the two major cellular membranes coordinate the intake and export of each plant nutrient, as well as other molecules that move in and out of the cytosol. Whether the mode of entry or exit is passive or active is mostly determined by ion concentrations in the cytosol. Thus, anions (such as NO_3^-, Cl^-, $H_2PO_4^-$, and SO_4^{-2}) are taken up actively because their concentrations are higher inside the cytosol. Boron in the form of boric acid can also be taken up passively, via membrane proteins as well as diffusion through the plasmalemma.

One exception to the anions moving into a cell passively occurs when there is a very low concentration of potassium ions outside the cell. In this instance, potassium can be actively transported into the cell. This says something about the importance of potassium to the operation of the plant cell. Other exceptions exist because the cytosol has lots of things in it, from large macromolecules to all sorts of metabolites as well as floating organelles, all of which can alter the ion concentration to be the opposite of the norm.

In any case, the anions accumulate on the cell walls of root cells, having been exchanged for hydrogen ions that are constantly being pumped out across the plasmalemma. Then they are actively transported into the cell. Anions, along with potassium ions (K^+), have the highest concentrations in the cell.

Sodium, magnesium, and calcium cations (Na^+, Mg^{+2}, and Ca^{+2}) are able to enter the cell by diffusion because their concentrations are lower. The opposite would be true about the export of these cations and any anions from the cytosol out of the cell. Calcium, sodium, and hydrogen (H^+) ions are all actively pumped out of the cell for use as electron sources for the active transport of anions inside. Calcium is also actively exported from the vacuole and from the cell. This may be to clear the boards for new signals.

Nitrogen has a positive ion, ammonium (NH_4^+), and a negative ion, nitrate (NO_3^-). Both are taken up actively, and this ensures the plant the ability to take up this crucial nutrient under different environmental conditions. Phosphorus only comes in the form of an anion, hydrogen phosphate ($H_2PO_4^-$), so its uptake requires active transport. Ensuring a supply of phosphorus in the cell is more important than the expenditure of energy required to import more. Iron, magnesium, zinc, copper, and manganese require active transport into the cytosol. Boron can enter the plant cell both actively and passively as an uncharged molecule, a major exception to the notion that nutrients enter plants only in ionic form. Suffice it to say this has to do with its chemistry when dissolved in water.

In sum, you have the answer to how plants eat. Nutrients in the form of charged molecules (ions) move into the cell walls and pass though the plasmalemma via transit protein channels, tunnels, and pumps. Some move passively via diffusion. Others are moved actively against the laws of diffusion as the result of some active transport mechanism. Note that it does not matter to the process if we gardeners team with microbes and let them do the work or if we apply artificial fertilizers. However, there are huge differences in the impacts to soil and the environment, which advocate the use of organic materials over synthetic ones.

TONOPLAST

So far we have been talking about the movement of nutrient ions into the plant through the plasmalemma of an epidermis cell. There is another crucial cell membrane, the tonoplast. This one is inside the cell, surrounding the vacuole and separating it from the cytosol.

> One of the beauties of a cell is that its membranes act in coordination

One of the beauties of a cell—and by now you should perceive many—is that its membranes act in coordination. It is this coordination that keeps the cytosol at a fairly constant pH of 7.2, so it doesn't damage organelles or drastically affect reactions. This is accomplished by pumping hydrogen ions out of the cytosol and into the vacuole, which is able to handle more acidity. The pH in the vacuole is low, usually around 5.5. The rest of the hydrogen ions are exported and accumulate on the outside of the plasmalemma, available for exchange and for use in pumping ions and molecules across the plasmalemma into the cell.

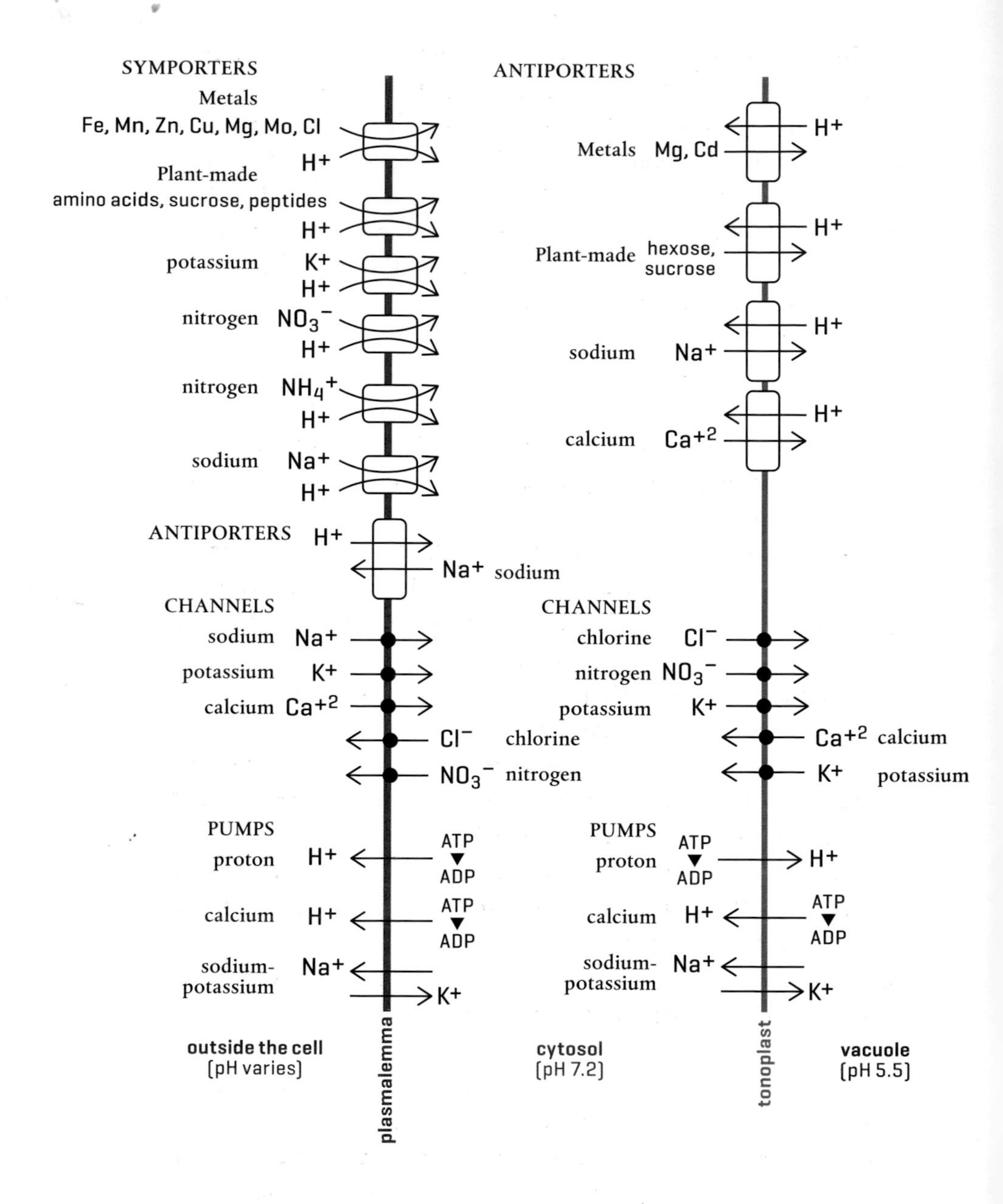

How some of the essential nutrients pass through the plasmalemma and tonoplast to reach the vacuole

The tonoplast, like the plasmalemma, tightly regulates what moves through based on all sorts of signals, such as pH, the concentrations of particular ions or proteins, light, heat, smoke, and salinity, with signals being sent between cells and from organelles to membranes. The membranes are just one part of the cell's incredible coordinated effort. Calcium, sodium, and hydrogen move into the tonoplast via active transport, whereas nitrate, potassium, and chlorine move into the vacuole via diffusion, a form of passive transport.

TRANSPORT PROTEINS

By now you should have a three-dimensional vision of the plasmalemma as a flowing wall with the viscosity of olive oil, punctuated with proteins of various kinds, shapes, and sizes that act like tunnels. Most nutrients enter and leave the cell via these transport proteins, which are designed very specifically for particular kinds of ions. Only those ions can cross the membrane using them, although often more than one kind of transport protein will accept the same nutrient ion. Indeed, you should have a similar vision of the tonoplast.

Because everything a cell needs to duplicate and to maintain itself has to be brought across a plasma membrane (at least one), there has to be an unbelievable number of these protein entrances. This is evidenced by the enormous amount of DNA plants dedicate to making each of these specific membrane transport proteins.

Although membrane proteins are very selective in what they will or can transport, they share enough similar characteristics to be grouped into three distinct categories: channels that molecules move through passively, protein pumps that provide energy for ions to actively cross membranes, and carriers that bind to specific nutrient ions.

► **Channel Membrane Proteins** Let's start with the passive mode of transport through a membrane, which occurs through channel proteins that span the membrane. They allow small molecules to move across the membrane without the input of energy. Again, these are very selective and only available for use by molecules that are a specific size, shape, and charge.

All of the anions diffuse out of a cell using channel proteins, whereas calcium (Ca^{+2}), potassium (K^+), chloride (Cl^-), and hydrogen (H^+) ions move into the cell via channel proteins. These nutrient ions only move through their own specific channel proteins. Much of a plant's water molecules are also transported across cellular membranes via special channel proteins called aquaporins.

Despite the fact that channel proteins don't employ energy to move ions, passive transport through them can be fast: 100 million molecules per second. This is faster than using any of the other methods of transport across a cell membrane. The regulatory nature of cell membranes dictates that channel proteins are not always open. In fact, they only open for a tiny fraction of a cell's life. The gates operate in response to a cellular or intercellular signal (such as a change in voltage, pH, or the presence or absence of a particular chemical). This allows for close regulation of materials that are capable of moving across a membrane with no input of energy and that might therefore flood the system.

► **Carrier Membrane Proteins** The second category of membrane transport proteins do not form a direct connection, a tunnel, all the way through the membrane, as do protein channels. Instead, carrier proteins require substances to bind to them for their movement across the membrane. These proteins are sometimes referred to as cotransporters, and

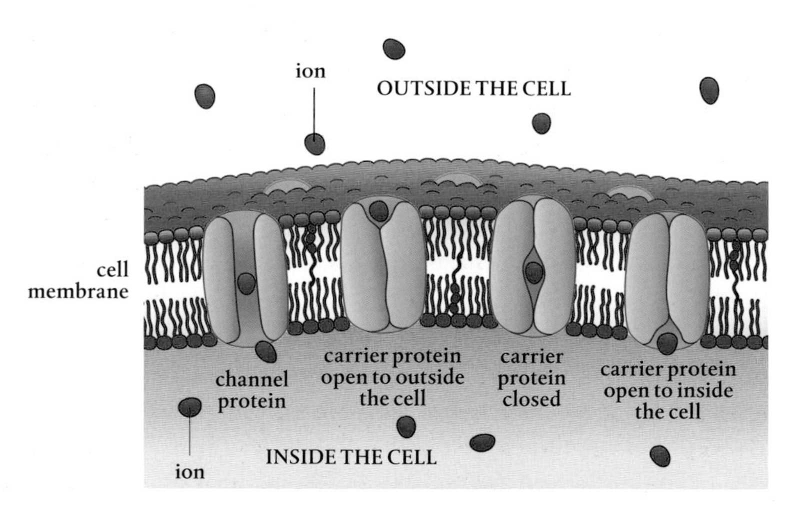

Carrier and channel proteins are designed to allow only specific ions or groups of ions across cell membranes. A considerable amount of cellular energy is devoted to making these transport proteins.

they can move nutrient ions actively or passively. The binding of the ion to the carrier protein causes the protein to change shape and in doing so to deliver its cargo to the other side of the membrane. Because the molecule or ion transported must bind to a specific site on the carrier protein, there is a tremendous amount of specificity dictating what molecules will bind to any given carrier protein.

Carrier proteins operate somewhat like enzymes. They only bind to specific nutrient ions to which they have a very strong attraction. When the nutrient is delivered, the protein returns to its original shape, with its binding electrons reverting back to outer side of the membrane. The bonding and detaching operates due to valence electron pairing without any energy input. This is much the way enzymes attach to and detach from their substrates. It's all chemistry, and chemistry is how things bond together.

There are three kinds of carrier proteins. Uniporters move their cargo down its concentration gradient, which is a form of facilitative transport. You might think this assistance would make it faster than moving through a channel, but that isn't the case. A carrier protein can only transport about 100 to 1000 ions every minute, way short of the tens of millions and more using the channels. It's the difference between traveling by ship or by airliner.

The other two carrier proteins, antiporters and symporters, actively transport ions as they move them against the energy gradient. Instead of using ATP to supply energy, these proteins use the electrical and chemical concentration gradients of other substances, usually hydrogen ions. The transported ion moves in conjunction with the other substance: symporters in the same direction like a booster rocket and payload, and antiporters in opposite directions like a catapult. Nitrate, iron, zinc, molybdenum, copper, magnesium, manganese, and chlorine move into the cell via carrier proteins.

► **Protein Pumps** There are a few kinds of protein pumps. The first are proton pumps or ATPases. They don't move nutrients, but rather hydrogen ions (which are essentially protons) generated by oxidation during cellular activity. As protons are pumped across the membrane, they accumulate outside the cell and create an energy gradient. Because they are charged particles, the protons can't diffuse through the membrane to get back into the cell. Instead, they travel down special enzymes, ATPases, embedded in the membrane. The ATPases use the movement of the protons to make ATP.

A second kind of pump uses ATP. The sodium-potassium pump works by binding sodium ions to the embedded protein and using ATP to change its shape and move the sodium to the outside of the membrane. After the release of the sodium ions, the protein's new shape is attractive to potassium ions, which bind to it, causing another transformation in shape and depositing the potassium inside the cell. This exposes the shape that is attractive to sodium, and the cycle begins again. These pumps move potassium and sodium against their concentration gradients at the expense of using cellular energy.

A third important pump is the calcium pump. Calcium ions have to be maintained in low concentrations to maintain the sensitivity needed for participation in signaling. Thus, calcium ions are pumped out of the cell to keep them clear, so to speak. Once a signal is sent, the calcium ions involved must be removed so that the next ones can be sent. These pumps use ATP to move the calcium in a fashion similar to the sodium-potassium pump.

MOVING NUTRIENTS THROUGH PLANTS

Nutrient ions travel along the same pathways as water into the xylem. The apoplastic pathway through cell walls presents far fewer barriers to the ions than does the symplastic pathway, which is extremely selective insofar as ions are concerned.

What drives water with nutrients to flow into the plant? We've already covered mass flow and the transpiration pull of the xylem sap. Transpiration only works during the times the stomata are open. When these leaf pores are closed or when the evaporation rate is very slow, the sap accumulates in the xylem cells. This causes water to flow into the cells to dilute the sap (that is, osmosis occurs). Because the sap is in tight sieve tubes that can only stretch so far due to the properties of their walls, the internal pressure created pushes the xylem sap up the plant where there is less pressure. Sometimes this osmotic pressure is so strong that it pushes sap out of the tips of leaves, a process known as guttation. While effective, root pressure is not capable of pushing a column of sap very high; several meters is about the limit. It is the passive pull of transpiration that does most of the work in getting water and nutrient ions to the tops of tall plants.

Once inside a plant cell—once a nutrient ion has passed through the right transport protein—it moves from cell to cell through plasmodesmata in the symplastic pathway or it might move back out into the less restrictive apoplastic passageway, only on the other side of the Casparian barrier. Ion exchange to the outside of the cell occurs with hydrogen ions that are pumped out of the plasmalemma, which lines the inside of the cell wall, and that attach to the charged polysaccharides that line it. Ions then continue traveling the apoplastic pathway or are absorbed again, moving through the plasmalemma via the many points of entry.

Root pressure is not capable of pushing a column of sap very high; several meters is about the limit

THE XYLEM TRANSPORTS WATER AND NUTRIENTS

Once nutrient ions enter the root's symplastic pathway, the solution they're in is pulled from cell to cell by the movement of water. This water is part of the stream of molecules being pulled through the plant as a result of transpiration. To get into the xylem apparatus, however, the solution must move out of a cell and into the apoplastic pathway, because part of the xylem is dead and there is no symplastic pathway into it.

Movement into the apoplastic pathway happens as a result of diffusion, root pressure, and the forces involved in transpiration, as we have discussed. All sorts of specialized membrane proteins are put to work, moving ions and water across membranes and into the cell walls of the xylem. Some are proton pumps that help load the ions into the apoplast.

Once in the apoplast again, the nutrient ions travel with water into the dead xylem vessels as a result of root pressure or the tension created on the water column by transpiration. As you know, it isn't just water that moves into a plant or in the xylem, it is water with dissolved nutrient ions. Some of these ions move out of the xylem before getting to leaves. Others are transported to sites of photosynthesis and to sites where proteins, lipids, carbohydrates, and nucleotides are synthesized.

During spring, large sugar molecules that were stored in the roots all winter are converted into transportable types of sugar. These travel through the xylem up to where they are needed. It doesn't happen often, but when sugars move via the xylem they create the perfect conditions for strong osmotic pressure. Water seeks to dilute the sweetened sap. As a consequence, it flows into the xylem tubes and pushes the sap up and out of the xylem system. Thus, the xylem tissue, although mainly responsible for the uptake of nutrient ions, is also used to move stored nutrients and water during the spring to the sinks at apical meristem tissues aboveground. This fact shocks many gardeners who were taught that the xylem carries only water. In fact, it is the xylem sap that is converted into maple and birch syrups.

Other items transported in the xylem during spring include amino acids and peptides. How does this work without transpiration, since there are no leaves? For birch trees, the answer is the root pressure created by the osmotic movement of water due to the sugars. However, this is not the case for maples. By observing maple sap produced once the trees go through freeze and thaw cycles, someone figured out what was going on. There are gases in the maple's xylem sap, which probably come from the production of carbon dioxide during cellular respiration of the sugars. As temperatures cool at night, these gasses dissolve, which lowers the pressure in the xylem vessels. Water molecules from the next cell over move in to equalize the pressure. Because of those hydrogen bonds keeping water molecules connected, this results in water being pulled into the plant. At night in the spring, it may get cold enough for the water in the xylem tubes to freeze and the gases it contains to get trapped in the ice. When it starts to warm up in the morning, the gases in the ice expand, re-establishing and increasing pressure in the xylem. Because the xylem can only stretch so far, an internal pressure is created and this pushes the sap upward. Once leaves and buds form in later spring, the maple and birch xylem sap stops carrying sugars and the phloem takes over the job. Consider all this next time you sit down to a plate of sourdough pancakes (my Alaska roots, if you will, are showing again).

Syrup and spring aside, the concentration of nutrient ions in xylem sap varies depending on extracellular conditions. If it is hot, a lot of water evaporates from the stomata, so there is lots of transpiration going on. At these times, the concentration of ions tends to be lower because of the higher volume of water moving through the plant. In the spring, mineral uptake is usually higher than during the heat of the summer.

Certain hormones also travel through the xylem when necessary. Some of these hormones provide the chemical signals that trigger and regulate stomata to close when there isn't enough water entering the roots. Others are growth-related signals. The presence of just one or two hormone molecules in a cell can send a message. (Given all the things that go on in a cell, the communication system must be something that Steve Jobs at Apple would have envied.) All it takes is one or two molecules. Amazing!

How does the water get from the xylem into cells where it is needed? Again, tribute must be paid to water's special properties. It can flow through membranes into the apoplastic or symplastic pathways. If water molecules move by cells that are low in water, osmosis occurs and water flows into those cells. Of course, there is always an aquaporin around to move water molecules. The vascular tissue of a plant is arranged so as to ensure there is ample access to water and nutrient supplies.

TRANSPORT OF SUGARS, STARCHES, AND SYNTHESIZED PROTEINS

The amazing thing about plants is that they can take fourteen nutrients in ionic form, along with carbon, hydrogen, and oxygen, and make whatever they need. They do so by synthesizing and assembling organic molecules and compounds.

Some of these molecules can be delivered from cell to cell via the symplastic pathway without involving vascular tissue, as long as the distance is only a few cells. For longer distances, water moves the organic molecules made in the green parts of plants to the rest of the plant via the phloem system. The phloem sap is basically water with a lot of sugars in it, as well as proteins and other materials made by plant cells that need long-distance delivery.

Carbon for these organic compounds is rendered into a useable form by photosynthesis in the leaves. This element then must be transported to all the regions of the plant where it is needed to combine with molecules to form proteins, lipids, or whatever else the plant requires to sustain itself.

The phloem system is also used to deliver exudates that are released from the plant roots to attract beneficial microbes to the rhizosphere. These plant-synthesized compounds play a key role in plant nutrition. They include some familiar and not so familiar acids such as citric, oxalic, piscidic, tartaric, acetic, lactic, and malonic acids. These names may not mean much, but each of these different compounds is made within cells from nutrient ions (or disassembled parts thereof) and then released from the roots in response to the plant's need, often related to special nutrient uptake. Each of these acids has different impacts on the uptake of phosphorus because they have different abilities to dissolve the various phosphate minerals. This is also true of zinc, iron, manganese, and other nutrients. Plants have an amazing ability to take those seventeen nutrients and make whatever compounds they need to obtain particular nutrients when they need them.

Plants have an amazing ability to make whatever compounds they need to obtain particular nutrients when they need them

Root exudates are lipid-, carbohydrate-, and amino acid–based molecules, and they travel to the soil through the plant phloem. Because carbohydrates and amino acids are very hydrophilic, getting them through the plasmalemma requires transport proteins that use energy. Because lipids are nonpolar, they move easily through phospholipid membranes by diffusion if they are small. Lipids can be quite large, however, and may require facilitative or active transport.

What is in phloem sap depends on the species of plant, as well as the stage of growth, season, and time of day. Oddly enough, glucose, the end product of photosynthesis, is found in very low concentrations and sometimes is totally absent from phloem sap. Instead, some 90 percent of the compounds in the sap are sugars in more complex forms. Some species of plants have sugar-alcohols instead of sugars (sorbitol, for example).

The pH of phloem saps is usually alkaline, between 7.5 and 8.5. Perhaps this is because amino acids only account for up to 12 percent of phloem sap. In contrast, the saps found in vacuoles and in the xylem tend to be acidic. The exceptions are perennials in the spring, whose phloem sap is acidic.

While the role of the phloem sap in transporting sugars is well understood, modern techniques used to identify other constituents of phloem sap have revealed the presence of messenger RNA, along with other proteins. These are a part of a long-distance signaling system. The RNAs identified use the phloem to coordinate responses to virus attacks and to set up defenses. These levels of RNA in the sap also seem to be a response to stress and to be involved in deciding where nutrients are used, in silencing genes, and in regulating development.

Depending on species, there can be lots of nitrogen-based molecules in the sap for another reason: nitrogen is a much needed element. Up to half of the solids in the sap in new seedlings, for example, can be nitrogen-based molecules, as so much of cellular synthesis in young plants requires nitrogen. In autumn, the proteins in the leaves are broken down and nitrogen is moved to the stems and roots for storage, as are carbohydrates.

In addition, different phloem tubes contain differing concentrations of molecules. As you might expect, smaller leaf veins might only carry carbohydrates, whereas larger feeder veins accommodate the larger amino acids. ATP molecules, potassium ions, hormones, organic acids, and even viruses are transported in phloem sap.

All parts of the plant must compete for sucrose, and those parts that are growing also compete for proteins and hormones. Partitioning is the process of dividing up these goods, so to speak. Plants grow very precisely. The roots grow in a way that enables leaves to get the proper amount of nutrients. (This is why, as any good gardener knows, if a plant loses leaves, you don't want to give it as much water because the roots need to adjust to the lower number of leaves they are feeding.) Flowers and seeds develop only after there is ample support for the synthesis required to produce them.

Sugar is one of the signals that control partitioning. Its presence stimulates the growth of new cells in the root, for example. If leaves are growing, they photosynthesize more sugar, which goes to the root (the sink) and further stimulates growth. However, when there is too much sugar in leaves, photosynthesis slows and may even stop. If the roots don't call for sugar, it accumulates in the leaves and new cells are not added to roots, which slows the whole process. It is an amazing machine, the plant.

SUMMING UP

So there you have it. Now you know how plants eat and a bit of what they do with their food. You should understand the reason for learning about the plasmalemma and tonoplast, phospholipids, hydrogen bonds, the apoplastic and symplastic pathways, diffusion, osmosis, xylem, phloem, and the Casparian strip. Nutrients must be in ionic form so they can be actively transported across membranes. Because water is a polar molecule, ionic compounds can dissolve in water. What, when, and how nutrients are taken in has everything to do with the stage of plant growth. When you feed your plants (or when you feed the microbes that feed your plants), think about root hair cells. Consider the flow of water through the soil to and through the root hairs, into the plant, and into the xylem.

By now you are probably reading chemistry and botany like you have degrees in both. The plasmalemma! Plasmodesmata, aquaporins, and transport proteins that help ions across the phospholipid membrane, and a double one at that! It should all make some sense now. You have the words. You know the basic concept of how nutrients enter and move through plants. The difficult study is done. Congratulations. Armed with this knowledge and understanding of how plants eat, we can now look a bit more closely at what plants do with these nutrients.

KEY POINTS

- **Interception, mass flow, and diffusion** are the three ways soil nutrients move to plants' roots.
- **The nitrogen-fixing** bacteria *Rhizobia* and *Frankia* and mycorrhizal fungi deliver plant nutrients to roots for uptake.
- **The plasmalemma** regulates movement in and out of the cell.
- **Passive transport** occurs when molecules cross cellular membranes without the addition of external energy as a result of concentration differences on either side of the membrane.

- **Active transport** occurs when the cell uses energy to move molecules and ions across the membrane against their energy gradient.
- **The tonoplast and plasmalemma** membranes regulate the contents of the cell, in particular maintaining the pH of the cytosol around neutral to prevent damage to organelles.
- **Nutrient-specific transport proteins** are embedded in the plasmalemma, providing the mechanism for ions and large molecules to move across the membrane.
- **Channel carriers** are proteins that act as tunnels for the passive transport of molecules.
- **Carrier proteins** bind to their cargo and move it across the membrane either passively or actively.
- **Protein pumps** use energy to pump ions that help move other ions across the membrane.
- **Aquaporins** are channel proteins that transport only water molecules.
- **Once inside the plant,** nutrients move with water as a result of transpiration. They follow the symplastic or the apoplastic pathway until they enter the xylem and are transported up and throughout the plant.
- **The xylem system** is unidirectional, moving water and dissolved nutrients up into the plant. In the spring, sugars and other materials stored in the roots move back up into the plant via the xylem.
- **The phloem** is a bidirectional system that transports sugars, starches, and synthesized proteins.

SEVEN

The Molecules of Life

WHAT'S IN THIS CHAPTER

- **Carbohydrates,** such as the celluloses that make up cell walls, are built from smaller sugar molecules produced via photosynthesis.
- **Proteins** are constructed from various combinations of the twenty amino acids, which are nitrogen-based molecules.
- **Specialized proteins** are involved in moving molecules and nutrient ions across cell membranes.
- **Enzymes** are specialized proteins that are necessary to catalyze chemical reactions that occur in plant cells.
- **The building blocks** of lipids are fatty acids and glycerol.
- **Lipids,** which do not dissolve in water, are important components in membranes and other barriers that regulate what can enter a plant or a plant cell.
- **DNA and RNA,** two important nucleic acids, are huge molecules that hold and replicate the genetic code, which controls the chemical reactions used to make all of the compounds in a plant.

WHAT HAPPENS to all of the essential nutrients after they enter the plant? Because plants are autotrophic, these nutrient ions are used by the plant to make all of the molecules necessary for the plant to grow, to sustain and maintain itself, and to regenerate. All of this work is done in individual plant cells.

About 80 percent of the molecules in a plant are imported ions and water molecules. The remaining 20 percent of molecules are made in the plant using those ions and water. These new synthesized molecules are what make gardening such a fantastic hobby.

In the chemistry chapter we touched on the molecules of life, the term scientists use to describe these synthesized molecules. Again, there are four main groups: carbohydrates, proteins, lipids, and nucleic acids. With these, plant cells can make everything needed for growth and reproduction.

Let's do some imagination stretching. First, let's consider how many theoretical permutations a plant can make with the seventeen essential nutrients. The answer is 35,568,742,896,000, which is a truly mind-blowing number of possible molecules. Obviously, not all of these combinations are possible because of electron bonding restrictions, lack of electron pairs, molecular shape due to special bonds, and other topics we have covered. Still, a plant cell has 100 to 200 trillion atoms, so we know there is a huge number of compounds in a plant cell. And there needs to be. A DNA strand can have hundreds of thousands to millions of nucleotide pairs.

It does take imagination to acknowledge that these diminutive plant cells contain lots of molecules of all sizes and shapes. It is much more difficult to comprehend that these molecules are being made by the trillions every single second in any plant you see. Molecular bonds are what make the combination of atoms into molecules possible.

CARBOHYDRATES

Most people are familiar with carbohydrates for dietary reasons at least. These molecules are produced by plants using photosynthesis. Carbohydrates consist of carbon, oxygen, and hydrogen atoms and usually have a base formula of $C(H_2O)_n$—that is, a combination of carbon and water molecules. In plant cells, sugars and starches are the key synthesized carbohydrates.

Precisely how carbohydrates are linked results in groupings called monomers, dimers, or polymers, which are single-, double-, or multi-chained molecules, respectively. Glucose ($C_6H_{12}O_6$) is a monomer and is what plant and animal cells break down to release energy. If you change the shape of glucose by rearranging some of its bonds, it becomes fructose (also $C_6H_{12}O_6$). Many carbohydrates names end in *-ose*.

Two monomers link together to form a dimer. Linking two glucose molecules produces the disaccharide maltose ($C_{12}H_{22}O_{11}$), and a fructose and a glucose come together to make sucrose (also $C_{12}H_{22}O_{11}$).

Combining three or more monomers or one monomer and a dimer results in a multi-chain polysaccharide, such as starch ($[C_6H_{10}O_5]_n$, where *n* indicates the number of repetitions of this subunit), which stores energy in plant roots. In fact, starch is vital for energy in both plants and animals. Carbohydrates have lots of hydrogen bonds, which is one reason why they are such good energy and storage molecules. Plants capture the energy from sunlight, and then store that energy in the form of carbohydrate molecules.

Cellulose contains half of all the organic carbon in the Earth's biosphere. This polysaccharide carbohydrate (also $[C_6H_{10}O_5]_n$) is found in all plant cell walls. The *n* for cellulose can be 500 to 5000 subunits long. As noted, these carbohydrate molecules play a crucial structural and protective role for the plant.

PROTEINS

Protein molecules are also made up of carbon, oxygen, and hydrogen atoms plus a fourth element, nitrogen. Amino acids are the smallest parts of protein molecules, the monomers that make up proteins. Twenty amino acids exist in nature.

Each amino acid has the same basic structure: a central carbon atom to which is attached a hydrogen atom (H), an amino group (NH_2), and a carboxyl group (COOH), which is what makes these molecules acids. The remaining carbon bond (there are four, remember) links with a side chain group that gives the amino acid its uniqueness. With just twenty amino acids to work with, a plant cell can form long and short chains and lots of combinations to make a whole lot of different kinds of proteins—in fact, any protein needed by them, you, me, and almost everything else on Earth.

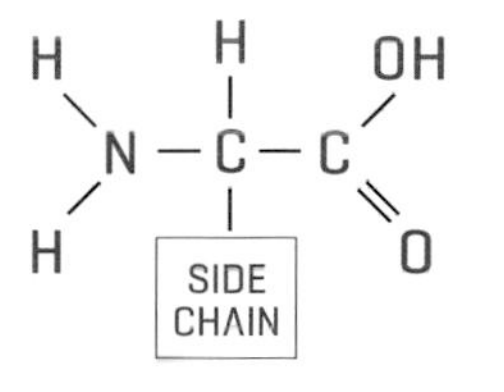

Plant proteins are made of combinations of twenty amino acids. Each amino acid has a central carbon (C) bound to a nitrogen-based amino group (NH_2), a carboxyl group (COOH), a hydrogen atom (H), and a variable side chain group.

Proteins are large molecules, 10,000 to 100,000 times larger than a single hydrogen atom. They are composed of peptides, shorter chains of amino acids that are linked by peptide bonds, which allow proteins to bend, fold, and twist. Actually, it would more accurate to note that the order of the amino acids and how they are bonded (especially the first three) is what determines a protein's shape. This folding is one reason so many can fit into a tiny cell and why they can move through cellular membranes and plasmodesmata. The folds and subsequent shapes of proteins can be affected by pH, temperature, and chemical signaling. Their shapes, along with the large size of protein molecules, make them well suited to serve as structural elements and to manipulate other molecules and atoms as enzymes. Carbohydrates may rule in the cell wall, but proteins are the basic building blocks of the cell itself. In fact, they play such an essential role that a tremendous amount of the DNA in a cell is dedicated to their synthesis.

All enzymes are proteins made by linking amino acids in a very specific order. Each plant cell produces thousands of different kinds of enzymes, each of which consists of hundreds to thousands of amino acid units linked together by peptide bonds. Once formed, enzymes take a unique shape as a result of their bonds and very long chains. These shapes allow them to carry out a specific reaction quickly. There are even enzymes to speed up the reactions that form new enzymes.

Enzymes can latch onto a specific molecule and either break it apart or make it bind to another. Try to imagine them folding and unfolding and twisting as they do their work. A single plant cell might have 10,000 different kinds of enzymes, and there may be 1 million copies of each in the cell. So, an enzyme is able to bump into the right molecules to fit together, which allows these reactions to happen so quickly. These protein enzymes are used to make lipids, carbohydrates, and nucleotides in the cell.

The names of enzymes usually end in *-ase*, as in lactase, the enzyme that breaks apart lactose. Proteases break down protein chains. Peptidases break peptide bonds to release amino acids. Lipids are broken down by lipases. (If there were a pill to reduce fat, it would surely be made of these.) Amylases break down starches, whose end products are sugars, and these are progressively broken down by maltase, lactase, and sucrase until the sugar monomer glucose is all that's left. I could fill this book with the complex names of enzymes like phosphoglucomutase and pyrophosphorylase.

Enzymes are extremely crucial to all cellular activities, be they metabolic or synthetic. There is no life without them. It should not come as a surprise that once an important enzyme cannot be replicated, a cell will die. Remember, life in a cell (life in general, actually) is only a series of chemical reactions. If you don't have enzymes to speed these up, you don't have life.

LIPIDS

Given their central role in the make-up of cellular membranes and the critical role these membranes play, it's no wonder lipids are considered molecules of life. As with carbohydrates, proteins, and nucleic acids, lipids are characterized by special configurations of chains of atoms. The building blocks of these lipid chains are fatty acids, which are composed of carbon, hydrogen, and oxygen atoms. Lipids are classified into fats, oils, waxes, glycolipids (sugar added), phospholipids, lipoproteins, steroids, terpenes, and carotenoids.

Lipids are important and necessary molecules of life because they do not react to the special properties of water. Because there are no charges on these long chains (that is, they are nonpolar), lipid molecules will not dissolve when put into water. This is why oil floats to the top of water. In fact, lipids stick together (coagulate) to avoid water. This property allows them to work well in membranes in and around a cell.

In addition to phospholipid membranes, lipids play an important role in the storage of energy. Their long chains have lots of hydrogen bonds that release lots of energy when broken. (Fats have more energy in them than any other molecule of life.) This means they are a great food source to directly provide energy, as well as a great source for storing it.

Steroids, composed of four carbon rings attached to a long hydrocarbon chain, are lipids. These molecules are necessary to make hormones,

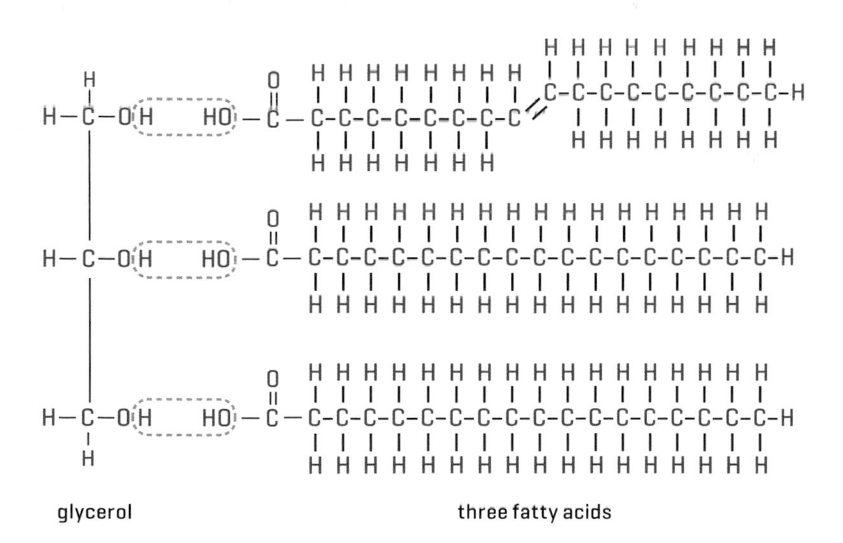

Lipids store energy and are used to make membranes. The basic structure of a lipid is long fatty acid chains bound to a glycerol or phosphate base.

which are signaling molecules produced by plant cells. Suberin is also a lipid. This is the waxy substance that clogs the cell walls in the Casparian strip at the end of the apoplastic pathway. In addition, the cutin on the outer layer of many plant epidermal cells is composed of lipids.

NUCLEIC ACIDS

Last, but not least, on the list of molecules of life are the nucleic acids, the two most famous of which are DNA and RNA. Carbohydrates, proteins, and lipids are useless unless they are put into some order and given some direction. The specific and replicable ordering of molecules is driven by DNA and RNA. Chemical reactions are what cells are about, and it is DNA and RNA that run those chemical reactions. One might argue where these originated, but there is not any argument that these two molecules enable life to replicate itself.

Nucleic acids also have building blocks, called nucleotides. Each nucleotide consists of a nitrogen base that has a phosphate and a sugar bonded to it. The proper names of DNA and RNA are deoxyribonucleic acid and ribonucleic acid. DNA has a double helix shape and RNA has a single helix, a result of the hydrogen bonds that connect the molecular strands and nucleotides in them. These two key nucleic acids differ only in that DNA has one less oxygen atom than RNA.

DNA is composed of four nucleotides: adenine, guanine, cytosine, and thymine. In RNA, thymine is replaced by uracil. Adenine always pairs with thymine (or uracil), and guanine always pairs with cytosine. Here is the key point of all life, at least on a secular, molecular level. DNA

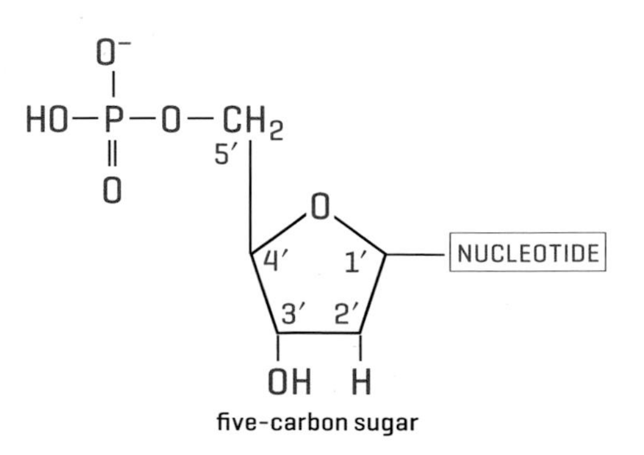

Combinations of nucleotides (nitrogen-based molecules) make up nucleic acids, whose general structure is one or more phosphate groups, a five-carbon sugar (deoxyribose or ribose), and a nucleotide.

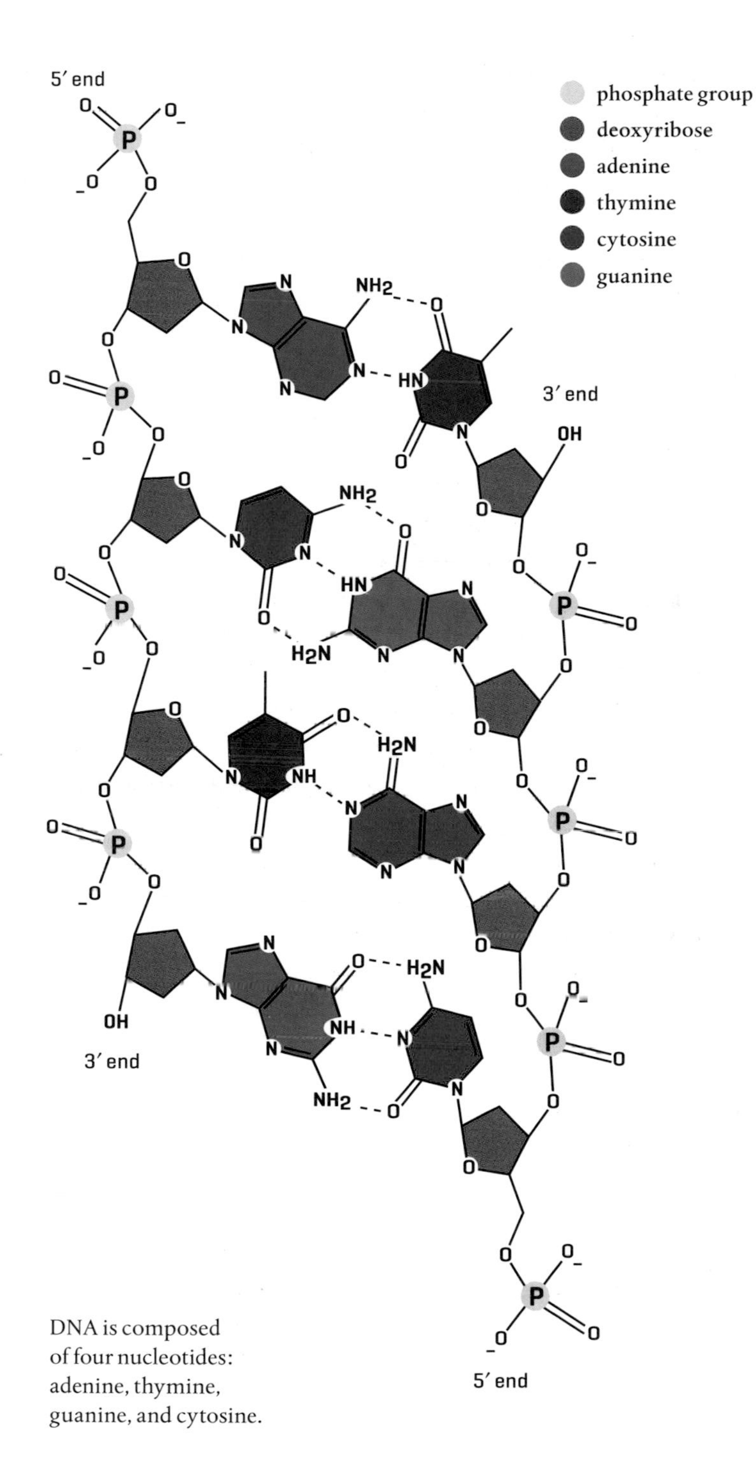

DNA is composed of four nucleotides: adenine, thymine, guanine, and cytosine.

is nothing more than a double strand of molecules. Put everything else aside because it is as simple as that.

Nucleic acids are huge molecules. The DNA of the bacterium *Escherichia coli*, for example, has over 4 million pairs of nucleotides. All this DNA is tightly folded, of course; otherwise the molecule would be more than 1 millimeter (0.04 inch) long. Human DNA comprises 3 billion pairs of nucleotides. There's a wide range in the numbers of nucleotides in the DNA of plants. *Fritillaria assyriaca* has 130 billion pairs, whereas *Populus trichocarpa* only has 480 million pairs. Even at the lower number, it's amazing to think that each of these molecules is contained within a nucleus.

RNA is a single strand of nucleotides. It looks like half a ladder, one cut down the middle. That way it can match with another half or make one that is a reverse duplicate and fits. There are several different kinds of RNA. Messenger RNA (mRNA) transcribes the DNA pattern in the nucleus and then travels to one of the numerous ribosomes on the endoplasmic reticulum or floating free in the cytosol. Transfer RNA (tRNA) gathers the amino acids coded by the mRNA molecule from the cytosol. tRNA uses ATP to attach to the amino acids and take them to a ribosome. Tracks in the ribosome serve as guides to hold units in place so the newly placed amino acids remain in the proper sequence. Once finished, the constructed protein is released from the ribosome along with the mRNA, and the process starts all over again.

Each of the proteins in a cell has a unique sequence of amino acids. The constituent amino acids are lined up by RNA in ribosomes to be linked into polypeptide chains. Hundreds of RNA nucleotides and hundreds of enzymes are involved in assembling each protein in accordance with instructions copied from DNA. It is amazing to consider there are millions upon millions of proteins in every single cell of a plant, and each one is constructed on site.

NUTRIENTS AND THE MOLECULES OF LIFE

Let's tie in the molecules of life to the nutrients that plants take up. First, let's consider the elemental composition of plant cell protoplasm, the stuff that's in the nucleus and cytoplasm of a cell. The most abundant element is oxygen at 65 percent. The next highest is carbon at 18 percent, followed by hydrogen at 10 percent. Surprisingly, nitrogen accounts for only 3 percent of a plant cell, just a bit higher than calcium at 2 percent and phosphorus at 1 percent. Potassium, sulfur, chlorine, magnesium, and iron together account for only 0.9 percent, and zinc, boron, cobalt, molybdenum, copper, iodine, nickel, and manganese make up the remaining 0.1 percent.

These essential nutrients are used to build the molecules of life—not only those in the protoplasm but also in the complex protein- and carbohydrate-studded plasmalemma and its supportive cell wall. Proteins account for 7 to 10 percent of the weight of a cell. You might expect this, given how many enzymes there are in the cytoplasm and protein transporters in membranes. Carbohydrates weigh in at 2 or 3 percent and are mostly located in the liquid part of the cytoplasm (the hyaloplasm), cell wall, vacuole, and as storage molecules. Lipids, present in the membranes of cells and the hyaloplasm, account for 1 to 2 percent. Nucleotides, RNA, and DNA are located in the nucleus, mitochondria, chloroplasts, and cytoplasm and account for about 1 percent. The rest, between about 85 and 90 percent of a cell's weight, is water. So, add water to these four kinds of molecules, and you have life!

Proteins account for 7 to 10 percent of the weight of a cell

► **The Assimilation of Nitrogen** I won't take you on the journey to see how each of the essential nutrients becomes part of the molecules of life or are used to otherwise regulate their construction. However, because nitrogen is such an important nutrient, it makes sense to at least briefly follow its path once it passes through the plasmalemma.

Again, nitrogen is assimilated into plants in two forms, as ammonium (NH_4^+) and nitrate (NO_3^-). When a plant takes in NH_4^+ and it enters into the cytoplasm of a cell, one of the H^+ in its ionic structure quickly combines with an OH^- typically floating around due to the relatively high pH. The result is a water molecule and an ammonia molecule (NH_3). This reaction changes the pH of the cytoplasm by decreasing the OH^- concentration (that is, increasing the H^+), which can quickly create toxic conditions and mess with the transport of electrons needed for photosynthesis and respiration. In order not to have this happen, ammonia is either converted into organic molecules or transported to a vacuole, where the conditions are acidic, meaning there are plenty of H^+ to return it to ammonium and thus a nontoxic state.

Nitrate is also converted into organic molecules. At first, this typically occurs in the roots, but as more and more nitrate accumulates, it is moved to shoot cells for assimilation. The conversion to organic molecules or assimilation causes NO_3^- to become nitrite (NO_2^-), which is toxic to plant cells. Nitrite molecules formed in roots are moved to plastids, and those formed in leaves are moved to chloroplasts. In each of these organelles, there is the enzyme nitrite reductase, which converts the nitrite to ammonium. This is then used to make glutamine, which is ultimately converted to glutamate. At this point nitrogen is available for use in different amino acids.

SUMMING UP

Rarely do gardeners stop to wonder at the fact that plants make all the molecules of life from just seventeen elements that are put together as a result of lots and lots of enzymatic activity, RNA, and DNA. It would take a college education to just scratch the surface of the various kinds of substances made in plants, not to mention understanding how they work (or how to pronounce their names).

Carbon, hydrogen, and oxygen atoms provide structure. Nitrogen, sulfur, and phosphorus atoms have special bonds capable of making uniquely shaped molecules, as well as storing and transferring energy. Potassium, calcium, and magnesium atoms are carriers and regulators of how the other nutrients get into the system. The other elements are used to make unique enzymes and other proteins or to assist in chemical reactions making various molecules.

In the nucleus are nucleic acids, which hold the instructions for how and when to put each and every single one of these atoms together into molecules. These instructions and the essential elements are used for absolutely everything from energy and structure to maintenance and growth. It is amazing that so few building blocks could result in such a wide diversity of the wondrous and complex organisms that plants are.

KEY POINTS

- **Carbohydrate molecules** consist of carbon, oxygen, and hydrogen and form large, repeating chains with lots of hydrogen bonds that store energy converted from sunlight.
- **Proteins are large molecules** made up of amino acids.
- **There are about 100,000** different kinds of enzymes, special proteins that speed up cellular chemical reactions. Lack of key enzymes will cause cellular death.
- **Lipids are composed** of fatty acids. They are important in membranes because they do not react with water. Lipids also have many hydrogen bonds that make them ideal for storing energy.
- **Nucleic acids** are made up of nucleotides and include DNA and RNAs.
- **Around 85 to 90 percent** of a cell is water. Proteins account for 7 to 10 percent; carbohydrates for 2 to 3 percent; lipids for 1 percent; and RNA, DNA, and the other nucleic acids for about 1 percent.
- **Nitrogen enters** the cell as ammonium or nitrate and is either converted to an organic molecule or moved to the vacuole. Nitrates are ultimately converted to glutamate.

EIGHT The Importance of Soil Testing

WHAT'S IN THIS CHAPTER

- **The use of fertilizers** should be based on information that can only be obtained by having your soil tested.
- **A laboratory** can evaluate the levels of the essential nutrients in a soil sample.
- **Other soil tests** include those for salt content, organic matter content, pH, cation exchange capacity, texture, and the presence or absence of beneficial microbes.
- **To avoid contamination,** use a plastic container and a wooden or plastic tool when taking soil samples.
- **Find a laboratory** that will provide the right kind of information, presented in a format that you can understand and interpret.
- **Various government agencies** and universities maintain soil laboratories and offer soil testing.
- **It's important** to follow the recommendations that accompany soil test results.

THE MODERN concept of fertilizers is predicated on Justus Von Liebig's Law of the Minimum, which states that a plant's yield is limited by the most limited nutrient. In agricultural situations, it is usually nitrogen, phosphorus, or potassium that are the limiting nutrients and hence the focus on these elements. This led to the development of artificial manures, which are now called fertilizers (talk about an image change).

This is a book about how plants eat, so the reference to Von Liebig's law is appropriate. This is also a book about organic gardening, however, so I should also mention another law, this one enunciated by Sir Albert Howard, considered the pioneer of organic gardening. The Law of Return expresses the need to recycle plant and animal wastes in order to keep the system healthy and producing humus. Historically, these two laws of agriculture have been pitted against each other, but when it comes to gardening, they belong together. Simply put, if you apply Sir Howard's Law of Return to the strictest extent (as you might with perennials and ornamental trees), then none of the nutrients in your garden should be limited, as per Von Liebig's law. When a plant dies, it is decayed by the soil food web, and the fourteen mineral nutrients that it contained are returned to the soil. A seed of that plant grows in the same soil and eats the same nutrients that were in its parent. No one needs to fertilize California's redwoods because they feed themselves in this way.

Farmers break the law when they take manures off the property and when they sell the livestock that ate the plants growing on the farm. Ah, but you are a gardener. Removal of animal waste is not the problem. You grow a cabbage, and at the end of the season you take it—and all the nutrients it has taken from the soil—into your house to be eaten. Hopefully, you leave the stem and roots, as per Sir Howard. But unless something is done to that soil, in future gardening seasons you will run up against the Law of the Minimum. Even removing weeds from the garden depletes the soil of nutrients. If conditions such as soil compaction, water, and temperature are in perfect shape (the goal of the gardener), it is breaking the Law of Return that results in the need to fertilize. (Having once been a government lawyer, I'll call this the Von Liebig–Howard Regulation.)

WHY FERTILIZE?

The apple really doesn't fall far from the tree. It drops, rots, and decays right where it will do the most good for the tree that produced it. Eventually, with the help of the soil food web, the apple returns the fourteen mineral nutrients that make up every single cell in it and its parent tree back into the soil, just over the tree's root system. So do the leaves that fall. When a gardener removes the apples, rakes the fallen leaves, or picks up the dead limbs and detritus after a storm, those would-be recycled nutrients are removed from the system. Eventually, enough are removed that if a plant is to survive, nutrients must be added.

There are two points to take away here. First, many plants in a yard really don't need additional input of nutrients and never will. They feed themselves. Ornamental trees and shrubs, for example, virtually never need additional nutrients provided that the soil food web is maintained and the Law of Return is applied. As long as the flowers are not harvested, most perennials don't need fertilizer, either.

Plants don't need fertilizer? This may sound contradictory to what you have been taught as a gardener, but this is science versus advertising. The rule should be obvious. Plants only need fertilizing when their soils become deficient in one or more of the fourteen mineral nutrients, regardless of what those ads say.

The second point is that once the system is out of balance because you ignored the Law of Return, you need to return the missing nutrients. Unless you do, your plants won't thrive, defeating the purpose of gardening. Fortunately, it is not difficult to do so.

GET YOUR SOIL TESTED

There is only one solution to making sure you are not breaking the Von Liebig–Howard Regulation: have your soil tested. Gardening interrupts the nutrient balance because we remove plants or parts of them from the system, and we often introduce plants into soils that can't support their particular nutrient demands unless additional nutrients are provided. The answer is not to just pile on fertilizer. In the majority of cases, gardeners simply assume fertilizer is needed and toss it on (organic as well as artificial) as a prophylactic measure. This practice is wasteful, and it can result in environmentally damaging runoff. The take-home message here is to test your soils before applying anything.

Knowing which nutrients are in the soil and which are missing is the only logical way to know what needs to be replenished. If you can get an idea of how much of a particular nutrient is in the soil, then you can also figure out how much more really needs to be added. This seems like such a simple concept, yet most gardeners have never had their soil tested. Frankly, until researching this book, I never tested any of the soils in my gardens or greenhouses for the essential nutrients.

Because other soil factors affect the availability of nutrients, these should be tested for as well. The same laboratory that tests for nutrient deficiencies can conduct these other tests using the same sample, so this makes a lot of sense as well. A soil's pH and its cation exchange capacity (CEC) tell a lot about the availability of nutrients. Many soil testing laboratories can also measure the content of organic matter in the soil, which is particularly important for those of us who garden without synthetic chemicals. Finally, to know how the soil food web, the organisms that make most natural fertilizers available to plants, is doing, biological tests are in order, although often these have to be done by a different laboratory.

► **Finding a Laboratory** It makes sense to find a soil testing laboratory in your local area, if possible. A local testing facility will have more experience testing and working with the kinds of soils in your garden and should know what inputs have corrected similar problems. They can also tailor the advice for collecting samples because they know what to expect. The Internet makes it easy to locate a nearby soil testing laboratory, as well as to check up on its reputation. What counts most

SOIL TESTING

Most books on gardening recommend getting soil tested, yet very few gardeners do so. This is a big mistake. How else are you going to know what your soils lack in terms of plant nutrients? Testing is the only way to get this information. Getting soil tested should be a regular part of gardening, not just something that gardeners pay lip service to.

It isn't difficult to collect the test samples. Very little soil is needed and a simple wooden kitchen spoon and a clean plastic bag are the only equipment you need. Brass, bronze, and galvanized tools and containers should not be used, because these might contaminate the samples with copper or zinc. Most American state colleges and universities have a soil testing laboratory. But no matter where you garden, the nearest soil testing laboratory is only an Internet search away.

Only a small amount of soil is needed for a soil test.

is consistency between tests, so that whatever laboratory you choose, you'll be able to compare test results from one year to the next.

A key factor that distinguishes laboratories is their reports. It's important to find a laboratory that will provide you with the right kind of information, presented in a format that you can understand and interpret. You should be able to get sample reports from several laboratories to help you make a determination as to the one or ones that fit your needs.

Don't overlook government agencies. In the United States, for example, offices of the Cooperative Extension Service, Soil and Water Conservation Districts, the Natural Resources Conservation Service, and the Farm Services Agency will test your soil or refer you to a laboratory that will. Most countries have similar agencies. In addition, many universities maintain soil laboratories and offer soil testing. Because universities are not normally in the business of selling products or nutrient systems, they provide straightforward test results.

Home testing kits are available, but most only test for nitrogen, phosphorus, and potassium and not the other nutrients. There are fourteen essential mineral nutrients, not just three, so these kits are of limited value. Similarly, home pH tests are not always reliable.

Most important, all good soil tests reports come with specific suggestions of how to correct any nutrient deficiencies or excesses. This is what makes the tests worth their minor cost. Most gardeners spend money on seeds and plants as a matter of course, and all invest lots of time tending to plants. Therefore, it makes sense to spend a little extra money (tests can cost as little as ten dollars) to test the soil that supports a garden so that the other investments in time and money are not wasted.

Although each laboratory has its own system for testing soils, essentially they all use the same process: soil is mixed in water, a chemical is added to free up the nutrients, and they are measured. The differences come about mostly in the type of chemicals added to release the nutrients. Some facilities use very strong acids, some use weaker acid solutions, and others use a simple water extract. Each system has its own merits, but some insist that using strong acids does not result in a real measure of available nutrients because those acids are not found in the soil. Some soil testing laboratories use patented methods, while others follow the teachings of famous agricultural professors who've lent their names to the test procedures.

If some of your plants are showing signs of nutrient deficiencies, you may want to test plant tissue in addition to testing the soil. Frankly, as much as we would like to believe that visual observations can tell us what is missing in a plant's diet, the truth is that many problems will cause the same symptoms to appear in a leaf, for example. The only way to really know what a plant suffers from is to test the plant tissue. Information is power.

I have one final point regarding soil testing and choosing a soil laboratory. I am a firm believer in knowing as much as you can about the food web in your soils, as well as the nutrients they contain. In fact, some would argue that this is more important information than knowing the soil chemistry, although both are important.

The soil food web should be cycling the nutrients in your soil into inorganic, ionic forms that can be used by plants. Knowing the status of the food web will help you understand if that part of your system needs adjustment to help ensure your plants receive the necessary nutrients on a timely basis. We know that mycorrhizal fungi play an important role in nutrient uptake. A test may be in order to determine their presence and population health, for example.

Biological testing is not the same as a chemical testing, and these laboratories have an even greater variety of ways to report their results, which are often not comparable between laboratories. Therefore, it is important to shop around and get sample test reports. Biological testing may require a laboratory different from one that tests for mineral nutrients.

Again, you can use the Internet to help you make a decision. The laboratories' web pages often show sample reports as well as suggested procedures for gathering the soil. Whatever laboratory you choose, once you're comfortable with their system, stick with it. This will enable you to have consistency in test results so that you can compare them to see which way your efforts are trending.

► **Soil Sampling** The soil testing laboratory will give you specific instructions for collecting, packing, and shipping samples. In general, however, you will need to take several samples from different areas (such as vegetable gardens, annual flower beds, lawns, and orchards). Also, take samples in areas that are problematic as well as those that do exceptionally well.

A soil test is only as good as the samples. It's important not to contaminate the soil samples being collected for testing. Brass, bronze, and galvanized tools and containers should not be used. These can release copper or zinc into the samples, messing up these values. Because copper and zinc are reactive cations, they can distort other nutrient calculations as well. Instead, use a plastic container and a wooden or plastic tool.

If you are collecting soil samples for a biological test, make sure the container and tools have been sterilized. (I once got a pickle-based set of test results, because I failed to follow that rule and used an old pickle jar.) And don't touch the soil either. If you need to clean the soil of sticks and stones, run the sample over with a rolling pin while the soil is in a plastic baggie, shake the bag to separate fines, and then pour out the unwanted stuff.

The soil testing laboratory will also specify the depth of the soil sample. For gardens, it is a good idea to take a sample down to 6 inches (15 cm). For shrubs and trees, go 4 to 6 inches (10 to 15 cm) deep and take samples from under the drip line. For lawns, 4 inches (10 cm) from the base of the plant down should be sufficient. Take several samples in each area. Although it sounds like a lot of samples, you can mix all the samples from one area together for a single test.

THE BASIC TESTS

There are several basic tests that you should make sure the laboratory conducts. First is a measurement of soil CEC, the ability of the soil to hold cations. In theory, this gives you what many agricultural laboratories call the "soil savings account," the total ability of the soil to hold positive cations. CEC ranges from 0 to 100, the latter being pure humus with lots of cations.

The CEC test is important for determining how much calcium, magnesium, potassium, and nitrogen in the form of ammonium is available in your soil. This is important because these elements almost always affect one another. Too much calcium can induce magnesium deficiency. Too much potassium has negative effects on magnesium uptake. Excessive magnesium isn't good for calcium availability. The CEC is also used to determine what to add to change pH when the problem isn't calcium related.

CEC tests can get very detailed. It's possible to measure how many cations are available to create acids, which can tie up some nutrients. A laboratory may also report the percent base saturation for individual cations in the soil. This tells you how many of the cation sites are taken up by particular cations. While this is not a test of the all the available nutrients, some laboratories use this number as the basis for their fertilizer program.

Closely related to the CEC is the soil organic matter content. (Remember, organic particles hold cations and exchange them for hydrogen ions produced by the plant root cells.) Soil organic matter is reported on a percentage basis, and most garden soils have around 4 to 8 percent. This is a good number to watch over the years, as it should trend upward if you add sufficient organic material each season. If it trends downward, you are not following the Law of Return or compensating for what you take out of the garden. Remember, organic matter doesn't mean fertilizer. Organic matter provides a place for the soil organisms to function, breaking down the fertilizers and making their contents available to plants. In addition, organic matter adsorbs nutrient cations for release to plant roots.

Next comes the pH test, which tells you the pH of the solution made by mixing your soil with water. The pH has a strong influence on nutrient availability, particularly that of phosphorus. The pH scale is logarithmic, meaning that moving one point makes a difference of ten times. So a pH of 8 is ten times more alkaline than a pH of 7. Again, the trend of the pH will tell a gardener much about his or her practices. If things get too acidic or basic, calcium or sulfur can be added to adjust the pH. However, the CEC, which usually increases with an increase of pH, also has an impact and is considered by a testing laboratory when making adjustment suggestions.

There isn't much mystery to the pH reading. Somewhere around neutral, a pH of 7, seems best. You can't change the pH of a soil instantly. It takes time. This is because the chemical reactions within the soil can continue to make the hydrogen ions that lower the pH. It is only after

A Sample Soil Test Report

Name: Homeowner		Sample date: April 9, 2012	
Lab number: 12345		Your sample number: 1	
Crop to be grown: garden		Sampling depth: 0 to 6 inches	
SOIL TEST RESULTS		INTERPRETATION	RECOMMENDATION
nitrate-N	12 lb/acre	low	3 lb. N/1000 sq. ft.
	6 ppm		
Olsen phosphorus	15 ppm	medium	2 lb. P_2O_5/1000 sq. ft.
potassium	192 ppm	medium	1 lb. K_2O/1000 sq. ft.
sulfate-S	15 ppm	high	—
boron	0.5 ppm	medium	0.02 lb. B/1000 sq. ft.
copper	1.7 ppm	very high	—
iron	47 ppm	very high	—
manganese	10 ppm	very high	—
zinc	1.3 ppm	high	—
soluble salts	0.3	low	—
organic matter	3.4%	medium	—
soil pH	7.7	medium–high	—
CEC	17.8	medium	—
soil texture	sandy loam		

the buffer acid ions are used up that the pH can be raised. This is the case when there are excess aluminum ions in the soil. They react with water and release hydrogen ions.

The bulk of a soil test report, however, is dedicated to the nutrient numbers. Sometimes these are divided between the macronutrients and micronutrients, but the results of each nutrient test show the availability of each nutrient and how much, if any, needs to be added. This is the most straightforward part of the report.

Most laboratories put a great deal of emphasis on nitrogen, which for most farmers and gardeners is the limiting nutrient. This can be reported in several different ways, including ammonium (sometimes expressed as ammonia) and nitrate availability. Although all of the essential mineral nutrients are important, plants use more nitrogen and it can be replenished readily, so laboratories tend to concentrate on it.

These tests were developed for farmers, whose living requires frequent soil testing, but we gardeners benefit.

Many laboratories cater to local conditions and may include special test results. For example, in areas where flooding by salt water is frequent, a sodium test might be conducted as a matter of course. This test is great for soils in northern climates that are affected by road and driveway salts, and a sodium test can be requested from all laboratories. By the same token, areas of heavy rainfall have different nutrient deficiencies than areas that don't get much rain. Even previous uses of the land can be a factor.

Many laboratories cater to local conditions and may include special test results

FOLLOW THROUGH

Obviously, it makes sense to follow the recommendations of the laboratory based on the soil test results. Study the material you receive and use your computer if you need more help. Even if they don't do testing, local agricultural agencies can provide lots of assistance once you are armed with soil test results. In addition, any decent laboratory will spend time with you on the phone or via email to answer questions pertaining to their tests, the results, or suggested remedial measures. If not, it isn't a good laboratory for you.

Make sure your laboratory knows the size of your garden, so they can make recommendations accordingly. If you are gardening in a small backyard, you don't need recommendations in pounds per acre or kilograms per hectare. Internet conversion calculators can also be used to bring things to the right scale.

It is also easiest if the laboratory you choose understands that you are an organic gardener or farmer and will provide recommendations using organic fertilizers. If not, you can access an organic fertilizer calculator via the Internet to convert recommended chemical fertilizers into organic ones. Oregon State University offers a good Excel spreadsheet for farmers and now has a separate one for gardeners. It is very easy to use (see Resources at the end of the book). Purdue University developed a similar free calculator for turf.

Finally, it's important to have a second soil test done even if you don't think your soils are deficient. The first soil test will establish a baseline, and a second test later in the first season or at the beginning of the next season will allow you to see where things are trending. A knowledgeable gardener gathers more than just great produce and flowers. Information is key.

FREQUENCY

For most home gardeners, and especially those who use organic practices, once you have the initial test results and at least a second test to see about trends, there is really no reason to test very often if you have been implementing the recommendations of your soil testing laboratory. You might consider a third check-up for annual, vegetable, and row crop garden soils 2 or 3 years after your last test. Lawns can go much longer without testing, 5 to 7 years. Soil around trees, shrubs, and display perennials left in place need only be tested if problems have started to develop.

The tests are not expensive and could be conducted each year. It is more a question of waiting long enough to see trending results. Still, there are times when it makes sense to test annually: if you are in a contest growing tomatoes or giant vegetables, for example, or if your income depends on growing the very best plants. Test annually if you drastically violate the Law of Return.

Discuss with your laboratory when appropriate follow-up tests should be taken—that is, when a test would show how your soils are being replenished or improving. Again, a local laboratory will have a lot of experience in this regard. You want to monitor trends, and sometimes these take a while to show up. Work with your laboratory.

Of course, as all gardeners know, the real test is how a plant is doing in the yard. If your plants are thriving, then you may have the right nutrients in your soil. Still, you now know how critical each and every one of the essential nutrients is. If an enzyme, for example, can't be synthesized because there isn't enough of a trace element, then something has gone or will go wrong with the plant. It makes sense to test your soils while things are working well, to ensure that there aren't any essential nutrients close to becoming that Von Liebig limiter.

SUMMING UP

After slogging through a bit of chemistry, cellular biology, and some botany in reading this book, you should understand that when the essential nutrients are not available to plants, they will not survive. It helps to know why, but from a gardening standpoint, it helps more to do something about it. All of this starts with soil testing, which should be a prerequisite to all other garden chores.

KEY POINTS

- **Gardeners violate** the Law of Return by taking plant material out of the garden, which does not allow it to decompose and return its nutrients back to the soil. Fertilizer is needed to return the nutrients removed.
- **The only way to know** what nutrients are missing from your soils is to have them tested.
- **Organic gardening** without chemicals is important to the soil structure, the soil food web, and the health of the environment. Look for a laboratory that provides organic recommendations.
- **When taking soil samples,** use wooden and plastic tools and containers to avoid contaminating the sample.
- **In addition to testing** for nutrients, test cation exchange capacity, pH, and any special local conditions.
- **Follow the recommendations** that accompany test results, and retest at the end of the season or the next year to discern trends.
- **You don't need** to test your soils every year unless you are seriously violating the Law of Return.

NINE

Factors Influencing Nutrient Availability

WHAT'S IN THIS CHAPTER

- **Temperature** affects soil microbes, water availability, and the chemical reactions that occur within plants, all of which influences nutrient uptake.
- **Certain pH levels** in the soil can lock up nutrients in chemical compounds, thus making them unavailable to plants.
- **Poor soil aeration** can cause compounds to build up to such levels that they are toxic to root cells and to beneficial soil microbes.
- **The mineral** and organic composition of soil affects the electrical charges on its particles, which, in turn, influences the soil's ability to exchange nutrient ions with roots.
- **Soil moisture** and a plant's growth stage also influence nutrient uptake.

IF SIMPLY ENSURING there was a sufficient and continuing supply of the essential nutrients to plants was all there was to gardening, we could all be prizewinners. Unfortunately, even if all the right plant nutrients are present in unlimited quantities, there are other factors that affect their availability to plants.

What makes a gardener a good one is understanding how plant nutrients work and how to supply them. However, what makes a better gardener is also being able to identify and deal with the special conditions required to ensure the essential plant nutrients are most efficiently taken up by plants.

TEMPERATURE

Plants don't grow (and some die) if temperatures are too cool or too hot. You can give nutrients to a plant under these conditions, but when the ground is frozen or when temperatures are above 95°F (35°C) or so, plants shut down and your efforts will be an exercise in futility. Plant cells need to be growing in order to take up nutrients, and the temperature has to be right for them to grow.

Temperature has a direct effect on the level of microbial activity in the soil. For example, when it is cool, as in the spring, there is less biological nitrogen cycling and mycorrhizal phosphorus. This is important because soil microbes provide more than half the nitrogen used in cultivated fields and gardens, and mycorrhizal fungi are responsible for much of a plant's phosphorus uptake. Although the microbes involved in providing nitrogen are active between 41°F (5°C) and 95°F (35°C), the highest production of useable nitrogen occurs in a temperature range between 75°F (24°C) and 95°F (35°C). (And now I know why gardening in Alaska can be so limited.) When it is cooler, the mycorrhizal activity key for bringing phosphorus, other nutrients, and water to the plant is slow and diffusion is slower.

Thus, phosphorus and nitrogen can be more difficult to get when temperatures are cool. Plants need to have ample supplies of these nutrients early in life or things don't turn out well. In early spring, in addition to taking steps to warm up garden soils, it is important that you ensure there are adequate supplies of nutrients so they are at least being taken up, even if slowly.

Temperature also affects water, and this, in turn, affects photosynthesis, respiration, and transpiration, all of which are intricately involved in nutrient uptake. High temperatures increase transpiration rates and the uptake of water and nutrient ions. Ions are transported in water, unless it is frozen. Because diffusion rates are temperature dependent, molecular movement slows down as temperatures cool. Stomata operation is temperature dependent, too, resulting in less transpiration in cooler temperatures and thus less water and nutrient movement.

Finally, enzymatic activity, of which there is plenty in a growing plant, is also temperature dependent. Generally, catalysts work better at warmer temperatures, although these systems break down above a certain threshold. Given that the growth and maintenance of a plant is catalyst driven, having enzymes work at the most efficient temperatures is the ideal.

There isn't a whole lot an outdoor gardener can do about temperatures, and that's exactly why gardeners resort to using greenhouses, cloches, cold frames, and heat-retaining mulches in cool climates and fans, shade-reflecting mulches, and the like in warm ones. What the gardener can do, however, is to plant the right kinds of plants. Sure, we all want a palm tree in our yards, but tropical plants don't do well outdoors in Anchorage, Alaska. By using plants that are suited for your climate, you'll lessen some of the impacts of temperature on nutrient uptake.

pH

Sooner or later most gardeners run across a chart that shows the impact of soil pH on the availability of essential plant nutrients. It appears in seed and plant catalogs, botany books, and even on some fertilizer packages. Most gardeners understand from it that soil pH (the measure of acidity of the soil) has an impact on the availability of certain nutrients. Few gardeners ever test their soil's pH, however, and even fewer ever try to really understand that ubiquitous chart.

The term pH stands for potential hydrogen. A soil's pH is simply a measurement of the number of hydrogen ions in the soil solution on a logarithmic scale from 1 (very acidic) to 10 (very alkaline), with neutral being 7. Water molecules are often split into hydrogen ions (H^+) and hydroxyl ions (OH^-) by chemical reactions in the soil. The hydroxyl ions go into the water solution, but the clay and organic matter in the soil attract the hydrogen ions, and the soil becomes more acidic. At some point the hydrogen ions take up all the available sites and nutrient

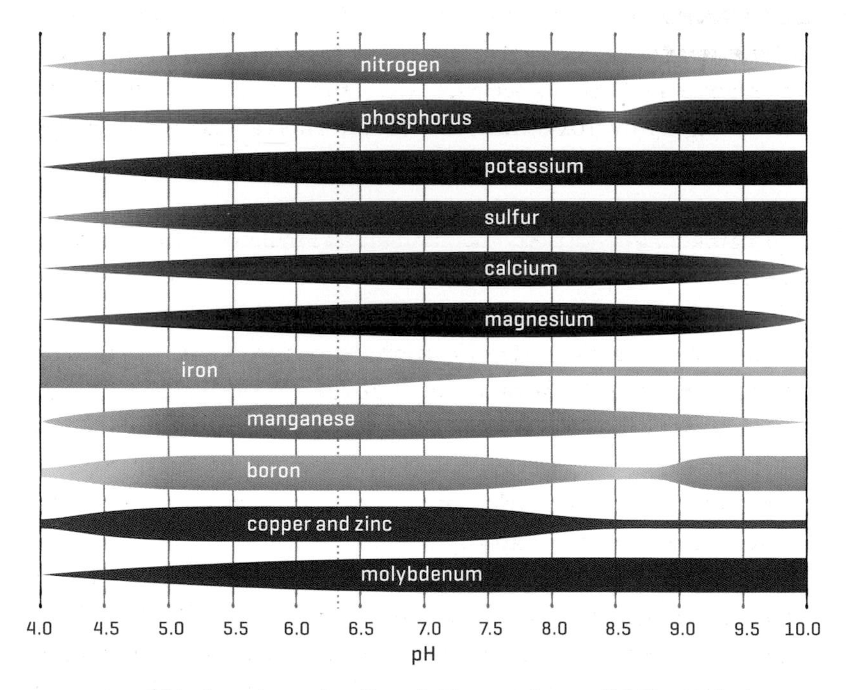

This chart shows the effect of pH on nutrient availability. Each element is most available within the pH range where the band is widest.

ions can't stick to the soil. The hydrogen ions go into solution only when there is no more capacity in the soil to hold them.

Problems with pH, frankly, plague chemical gardeners more often than natural and organic gardeners who use compost and are building humus in their soils. Most chemical gardeners wouldn't have a pH problem if they had their soils tested and followed recommendations. Anytime a gardener adds ammonium (NH_4^+), it is converted to nitrate (NO_3^-) by nitrogen-fixing bacteria. This results in three H^+ being added to the soil water, which lowers the pH. It doesn't matter if the ammonium comes from a natural source or a man-made one; the pH will be lowered.

Compost, on the other hand, has a very high percentage of organic matter, and so it generally has a pH in the range that is better suited for plant growth, 6.5 to 7.5. Adding and using organic materials greatly increase the ability of the soil to hold nutrient cations because the numbers of cation exchange sites increases. Such soil can hold lots of hydrogen ions and thus buffer the soil pH by absorbing them. Buffering is the ability of the soil to absorb acids and not change pH. Carbon dioxide forms a weak acid that breaks apart some of the water molecules in soil. In organic soils and soils with lots of clay, the hydrogen cation is tied up, leaving the hydroxyl anion in solution. This causes the percentage of hydrogen ions, the measure of acidity, to go down. Applications of organic matter generally increase the pH.

In general, soils in areas that have low rainfall have problems with too much sodium. Weathering leaves sodium carbonate in the soil. When it does rain, the sodium compounds in the soil react with some of the water molecules, releasing the hydrogen and hydroxyl ions. As the hydrogen ions attach to the clay and organic matter, the percentage of hydrogen ions goes down, and the pH goes up, making the solution more alkaline.

Most soil test reports will show the pH of the soil tested. This is the pH to which the plant is exposed. Often you can also get a measure of your soil's buffer pH. It is simple to then read that familiar pH chart and react accordingly (just like those nutrient molecules will). If you have a pH of 6, for example, you need to start to think about the uptake of molybdenum, magnesium, calcium, and phosphorus. If pH were the only test you had done, a reading of 6 would have you concerned about deficiencies of these mineral nutrients, as the pH chart depicts. This is one instance where adding more won't help. You have to raise the pH to make them available.

Finally, it is important to understand that in a natural system, the plant is in control. Remember all those hydrogen ions passed out of root hairs to adjust the pH inside the cell and those that serve in the proton pump systems sustaining active transport of nutrient ions into plants? The plant doesn't need a gardener to tell it when to operate these pumps. It adjusts its own pH inside the cell and outside in the soil because it has to in order to survive. This takes time. It also means the plant has to divert energy and resources that could be making better plants, which is why gardeners need to step in.

Plants synthesize and release exudates of various components that adjust the pH to where it should be

Plants synthesize and release exudates of various components that adjust the pH to where it should be. The plants do this with help from the type of microbial community these exudates attract and support (acidic exudates attract fungi and alkaline ones attract bacteria). The plant has a lot of control over the pH of the soil in which it grows. It's no wonder. The pH is important to how and even if nutrients are available to the plant.

The parent material of the soil has a lot to do with its pH. So does the water you apply to it. You should have the water tested when you first test your soil. It is doubtful that it will change much over time, so you probably don't need to test it again. In addition, water can increase the amount of ions available by weathering and dissolving minerals, freeing them as ions. It can also take ions away, literally, by washing them out to sea. So, the amount of water and the time of exposure to water are also important factors when it comes to pH, and this affects nutrient availability.

For example, when the pH gets below 6, phosphorus in the form HPO_4^{-2} becomes much less available because at lower pH this anion bonds more strongly to the cations of iron and aluminum. It is fixed, which in this case, unlike nitrogen's, means it is not usable. (What is with these chemists?) On the other side of the coin, if the pH gets above 7, too much becomes fixed by reacting with calcium and forming calcium phosphate. This renders phosphorus increasingly unavailable to plants in sufficient quantities.

The uptake of nitrogen is also greatly influenced by pH. First, acidity has a lot to do with the kind of nitrogen available to plants, ammonium or nitrate. One of the tenets of gardening with the soil food web is that if the plants are in the ground for less than a year, they generally prefer their nitrogen in nitrate form. Nitrifying bacteria, which produce nitrate, don't do well under acidic conditions. Fungi dominate in acidic conditions, however, and these produce the ammonium that perennials, trees, and shrubs prefer.

A high pH can result in the loss of nitrogen from the soil. There is only one H^+ difference between ammonium (NH_4^+) and it gaseous counterpart, ammonia (NH_3). The two are normally in a state of equilibrium, but if the pH goes from 7 to 8, then 10 percent of the ammonium becomes ammonia, which evaporates out of the soil. Other factors aid in this loss, namely temperature and moisture, so we can't just blame the pH. Still, it starts it all.

Of the macronutrients, only sulfur is not affected by pH. The micronutrients manganese, iron, copper, zinc, and boron become less available as pH increases. Molybdenum is more available when the pH is a bit alkaline and less so when things are acidic. As the pH decreases, aluminum binds with potassium and traps it between clay layers.

The wrong soil pH can cause some nutrient minerals to remain in solution, whereas others become insolvent. Iron, for example, is much more available at a pH of 6 than at a pH of 7. Normally plants will export more hydrogen ions into the soil to create the necessary acidic conditions to make iron more available. If the gardener adds limestone to raise the soil's pH, it can cause iron deficiencies. Nitrogen in the form of ammonium also increases the production of hydrogen ions from plants, thus aiding iron uptake. Nitrogen in nitrate form works in the opposite manner, causing the hydroxyl ion concentrations to increase, raising the pH, which makes iron less available.

When metal ions combine with organic compounds, a chelate is formed. This prevents the metals from participating in reactions, thus making them unavailable. This is particularly important when it comes to iron (Fe^{+2}). Lots of the essential nutrients have specific chelate counterparts, but they are not important because the nutrient is not affected within the normal range of pH. Fe^{+2} is, however.

The solution to an improper pH level is to correct it. The usual recommendation for acidic soils is to apply finely ground calcitic ($CaCO_3$) or dolomitic ($CaMg[CO_3]_2$) limestone, with calcitic lime used for soils with high magnesium levels and dolomitic lime for soils low in magnesium. The rate of application is directly related to the buffering pH of the soil, so that soil test is very important. The limestone binds with the hydrogen ions, and the end product of the reaction is carbonic acid (H_2CO_3), which further breaks down into carbon dioxide (CO_2) and water (H_2O). Correcting alkaline soils usually involves applications of sulfur and water, which becomes sulfuric acid. The acid reacts with the soil chemicals and becomes gypsum ($CaSO_4$).

While chemical gardeners have the most problems with pH, organic gardeners should occasionally test for pH, but normally need not worry too much about it. Just continually add compost and use mulches to increase the soil organic matter, which, in turn, increases the cation exchange capacity (CEC), food web populations, and thus its buffering capacity.

Finally, a gardener must be cognizant of the requirements of the plants in the garden. Some, most notably azaleas, blueberries, potatoes, pines and other conifers, hollies, and camellias, require acid soils. Obviously, these plants have evolved some mechanisms for dealing with tied up nutrients.

SOIL AERATION

The ability of your soil to maintain aerobic (oxygen-rich) conditions is another major factor that affects plant nutrients. Well-aerated soils have lots of microscopic pore spaces that allow for air and water exchange. This water replenishes the supply of nutrients to depleted root zones, and it carries nutrients up into plant roots. If soils are not well aerated, there can be less water and consequently lower root pressure and less mass flow and nutrient uptake.

In addition, carbon dioxide produced by cellular respiration in roots can build up in poorly aerated soils. Carbon dioxide chemically reacts with water to form an acid, and it often combines with organic matter to form cell-killing alcohols and fermented products that are not good for plant roots or some of the beneficial members of the soil food web. Well-aerated soils can absorb and then help release the carbon dioxide produced by cellular respiration in roots into the atmosphere. These soils also contain oxygen, which mixes with water and enters roots.

When soils are anaerobic (oxygen-poor), microbes that require oxygen often replace this element and use other nutrients instead. Thus, soil compaction affects the availability of iron, sulfur, and manganese, because the microbes use these nutrients, reducing amounts available for uptake by plants.

And, of course, soils need to have ample oxygen to sustain the free-living microbes that fix nitrogen, as well as the plants that house many of them. Moreover, the microbes that like anaerobic conditions include many that unfix nitrogen at the plant's (and gardener's) expense. Finally, mycorrhizal fungi, which are important for the uptake of phosphorus, nitrogen, copper, and other essential nutrients, require aerobic conditions.

Aside from oxygen, the uptake of potassium is the nutrient most affected by compacted soils. A whopping 50 percent reduction can occur in compacted soils. Because potassium is required for the regulation of carbon dioxide and water levels, which are both important for photosynthesis, it is no wonder plants become stunted in anaerobic soils.

Plug aeration is a great way to loosen up and get air into a compacted lawn soil. The plugs are left on the ground to decompose.

Finally, there is a fine balance between the proper amounts of air and water in soil pores. Although water is needed to deliver nutrient ions to plants, too much water displaces air, causing the soil to become anaerobic. Soils must be aerobic for plants to take up nutrients in the most efficient manner and quantities.

The compaction of soil is caused by too many things to list, but they range from too fine of a soil texture to layers in soils, too much rain or hose time, the wintertime ice hockey rink on the lawn, and the like. To correct the problem, aerate lawns and then add compost. Amend garden soil with compost and organic matter to increase the activity of the soil food web.

Unless you have lots of clay, rototilling is not a solution to compacted soil. It destroys soil structure and greatly reduces mycorrhizal fungi and many of the other beneficial soil food web organisms. This leads to more compacted soil. Instead, add lots of organic matter to the surface of gardens to fix the situation if you want the most efficient nutrient uptake and therefore best plants. When it comes to gardening in clay soils, rototilling to mix in organic matter may be required until enough is added for adequate drainage. Thereafter, surface applications should suffice.

Unless you have lots of clay, rototilling is not a solution to compacted soil

CATION AND ANION EXCHANGE CAPACITIES AND NUTRIENT MOBILITY

The ability of nutrients to move through soil to roots depends on the characteristics of the soil. Organic and clay particles in the soil have lots of negative charges on their surfaces that hold positively charged mineral nutrients. Theses nutrient cations can be exchanged with plant-produced cations. The number of cations that a soil is capable of holding is the cation exchange capacity (CEC).

Clay is made up of sheets of molecules, and some molecules hidden in the layers hold positive charges. When these become exposed, they attract anions that are exchanged with hydroxyl ions (OH^-) in the water solution. Likewise, the number of anions that a soil is capable of holding is its anion exchange capacity. Because most anions are already in the water solution and available to plants, however, this is not as important as the CEC.

Soils with a low CEC won't hold nutrients well, so the gardener needs to mete out nutrients over an extended period of time so they won't all leach away. If your soil has a good CEC, then you can dump in large amounts of nutrients and expect them to be held. If your soil is sandy, you may need to add compost full of organic matter and clay to increase the CEC of your soil. CEC may also influence the timing of fertilizer application. For example, you wouldn't want to put fertilizers down in autumn with low-CEC soils, because there would be nothing left by spring due to runoff. However, if your soil has a high CEC, amending it in autumn might be a good practice.

In short, CEC has a lot to do with the mobility of nutrients in soil. Assuming an adequate CEC, the anions chlorine, nitrate, molybdenum, and sulfur are mobile in soil. In contrast, the cations ammonium, calcium, copper, iron, magnesium, and manganese are much less mobile, depending on the amount of organic material and clay, which increase CEC and decreases their mobility. Nickel, phosphorus, potassium, and zinc are relatively immobile in soils. Mobile elements have to be replaced more frequently than the immobile ones because they are both readily taken up by plants and more quickly leach out of the soil.

SOIL MOISTURE

Obviously, the amount of water in the soil can have the ultimate impact on the availability of nutrients. Water influences pH, mass flow, and root pressure, all of which affect nutrient uptake.

As water moves out of the soil, it leaches away the nutrients dissolved in it. It can also wash away soil that contains nutrients, which is a major source of phosphorus loss. Of course, water can also increase nutrient availability by releasing nutrients, both chemically and by weathering. Once again, too little or too much water has a direct impact on populations of microbial symbionts. If there is too much water and anaerobic conditions develop, our *Rhizobia* and *Frankia* bacterial friends won't produce nitrogen. This can cause a decrease in the production of sugars in leaves and reduce the numbers of soil food web organisms that rely on exudates, including mycorrhizal fungi. The result is little or no nitrogen, phosphorus, copper, and other nutrients being delivered biologically to the plant.

TIMING

We've learned enough about plants to understand a bit better the timing of their needs. For example, if there is too little nitrogen at the end of a plant's growing cycle, that may not affect the fruit too much. But too little nitrogen at the beginning of the plant's life and you won't even get to experiment. Timing can mean as much as anything when it comes to supplying nutrients.

You should become familiar enough with each nutrient's impact on the development of plants to know what nutrients are especially needed at what times of the year. Key times to think about are seed germination, transplanting outdoors, flower development, and the first appearance of fruit.

During the initial growth stages of a plant—the first six weeks of annuals, row crops, and perennials each year—these plants grow like crazy and require nitrogen, phosphorus, potassium, zinc, iron, magnesium, manganese, copper, sulfur, and molybdenum in the greatest quantities. The same thing happens when a perennial plant starts up again or a tree, shrub, or houseplant starts a growth spurt. You should have a pretty fair idea of why: proteins are being synthesized, chloroplasts are needed, and cell membranes have to be working double duty to regulate all the materials needed. Once a plant gets to the reproduction stage and starts to develop flowers and fruits, they need ample boron and calcium. Boron is needed for pollen formation, and calcium is needed to produce the flowers.

Timing comes into play in other regards, too. In most gardens, the season starts when there are cool temperatures, with influences already discussed. The root system of annuals, for example, is small during the spring, yet it has to take up massive amounts of nutrients. One calculation has young roots picking up twenty times more nutrients per unit area than when the plant is finished with its initial growth stage and has a much bigger root system.

In general, annuals, row crops, and some perennials go through growth stages at the fastest rate. These are the plants that normally need fertilizing if the Law of Return is broken, the soil food web system isn't healthy, or the soil lacks one or more nutrients as shown by testing.

SOME FACTORS INFLUENCING INDIVIDUAL NUTRIENTS

Individual nutrients react differently to specific garden conditions. For example, boron, nitrogen in the form of nitrate, chlorine, and the other anions move to the plant in the soil water solution. Their uptake is a function of the amount and rate of water being transpired and is thus dependent on all the factors that keep transpiration happening. Thus, humidity is key, as are temperature, the time of day, and exposure to daylight or darkness.

Similarly, many of the essential nutrients are taken up by active transport. This is an energy-consuming activity, and thus all the factors that affect the ability of the roots to operate properly are important. There has to be the ability to make cellular energy directly, as in the production of ATP by proton pumps in membranes, or to get it by converting sugars made during photosynthesis.

In some instances, there is competition between ions for absorption sites on the roots. For example, sodium, magnesium, and potassium cations compete with calcium cations. So if you add too much of those nutrients, plants will take in less calcium.

Several factors influence the availability and uptake of iron. In addition to pH, the amount of organic matter in the soil is important. In times of soluble iron shortages, certain bacteria and fungi in soil produce exudates known as siderophores (as do a few plants). These bind to insoluble iron in a process known as chelation, making it much more soluble and much more mobile in the soil and thus more available to plants. In times of iron stress, plant exudates can increase these microbial populations. The presence of these microorganisms is one reason why compost, mulches, and manures are so helpful in keeping sufficient iron levels in plants and greatly reducing or eliminating the need to apply supplemental iron.

Iron also combines to varying degrees with calcium, copper, manganese, zinc, molybdenum, and phosphorous, making it unavailable to plants. In addition, iron is often bound up by siderophore-producing microbes, which chelate the iron for themselves, and plants just happen to benefit.

Factors also influence the uptake of nitrogen. Plants usually prefer one of the two types available, nitrate and ammonium. Up to 50 percent of a plant's nitrogen can come directly from nitrogen-fixing bacteria. They operate best at warmer temperatures and at a pH of above 7, with numbers diminishing as the soil becomes more acidic. In addition, ammonium is taken into roots and combined with sugars there to make proteins. These sugars are produced in the leaves and are transported to the root cells. Nitrate, on the other hand, is transported up from the roots to leaves and is there converted into ammonium. A lot of sugar is consumed in the production of leaves, flowers, and fruits during very active growth, such as on a warm day, and there may not be enough sugars to send down to the roots to combine with the ammonium to make proteins there. The ammonium is not being used efficiently and plants use less. The reverse would be true at lower temperatures, when ammonium is more readily converted in the roots.

Nitrogen in ammonium competes with calcium and magnesium ions, and less so with potassium ions. Apply too much ammonium, and you may end up with not enough of the other cations. This, of course, will have a negative impact on your plants. One more thing about nitrogen: we know that the release of hydrogen ions to get ammonium through the plasmalemma results in lowering the pH around the root zone, whereas taking up nitrate results in a higher pH. So the pH can be affected by the choice of nitrogen offered and used.

Finally, individual nutrients travel at different rates in the soil. This depends on the soil CEC and the amount of water available. Where these nutrients are placed by the gardener has a lot to do with their availability. As you would expect, the immobile nutrients need to be placed closer to seeds and roots to be effective. This is why phosphorus compounds are best placed in bands just under seeds and in the zone where roots will grow. It is also why some companies coat seeds with immobile nutrients.

SUMMING UP

There are lots of factors that influence the availability of nutrients, including chemical, biological, and environmental factors, and they create a complex matrix. This complexity is one more reason why gardeners should be getting soils tested and following recommendations. The testing facility considers all of these complicating factors in coming up with recommendations. Following them will make you a better gardener and gardening a better experience by getting your soils and these factors in balance.

KEY POINTS

- **Temperature influences** soil microbial activity, cellular enzymatic activity, and soil and cellular water flow.
- **Both high and low pH** ties up needed nutrients.
- **Nutrients** chemically influence each other at different pH levels. The recommendations of soil testing laboratories take all of these relationships into consideration.
- **Soil aeration** ensures proper microbial activity and water flow to deliver nutrients to plants.
- **Cation and anion** exchange capacities and nutrient mobility are affected by organic matter and the clay contents of soil.
- **Anions are mobile** in soil, whereas cation nutrients are much less mobile. The exceptions are nickel, phosphorus, potassium, and zinc, which are relatively immobile in soils.
- **Soil moisture** influences the presence of microbes, pH, and mass flow of water in the soil.
- **The right nutrients** need to be available during the stages of life when they are needed.

TEN What and When to Feed Plants

WHAT'S IN THIS CHAPTER

- **Healthy plants** result from a healthy soil food web.
- **Unlike chemical fertilizers,** natural fertilizers foster the health of the soil food web, which builds all-important soil structure.
- **Good natural sources** of nitrogen include bat guano, blood meal, corn gluten meal, and even human hair and urine.
- **Biofertilizers** are living organisms that are added to the soil to promote plant health. These include nitrogen-fixing bacteria, phosphate-solubilizing bacteria and fungi, and mycorrhizal fungi.
- **I provide** some organic fertilizer recipes designed for annuals, vegetables, lawns, perennials, trees, and shrubs and describe the best ways to apply these fertilizers.

NOW THAT YOU know how plants obtain the essential nutrients that are in fertilizers, it's time to discuss which are the best to use and when. No two gardens are alike, so the choice of fertilizers to use has to be an individual one based first on sound soil testing. The rest is up to the garden and the gardener. It is no longer acceptable to simply buy a package of fertilizer based on a picture of the plants you want to feed.

USE NATURAL FERTILIZERS, NOT SYNTHETIC

For starters, use natural fertilizers, not synthetic ones. Some would argue that it can't possibly make a difference to a plant if its nitrogen and other essential elements come from the remains of another plant or an animal or if the fertilizer is synthesized in a factory out of chemicals. These nutrients have to be in inorganic form for uptake. Synthetic fertilizers do work in the same way as natural ones once inside the plant. And, for the most part, the organic matter in natural fertilizers has to be reduced to inorganic compounds before nutrients can enter a plant. The big difference and the reason organics and natural fertilizers are better, however, is because they feed the soil food web, which makes soil structure. Inorganic fertilizers generally do not build or help sustain soils.

Chemically, the implication of using synthetic fertilizers is clear. They can tie up essential nutrients, quickly change the pH, and adversely affect osmosis because of their high concentrations. The ultimate repercussions are symptoms of poorly nourished plants. Synthetic fertilizers also have a negative impact on the soil biota. Once the soil food web organisms are affected, either directly or indirectly, the soils lose the structure these organisms create. Bacterial slime and fungal hyphae initially stick and weave soil particles together, creating pore spaces, reservoirs for air as well as water and places for the smaller organisms to hide from predators. Tunnels and burrows further increase the air- and water-holding capabilities of the soil. Dead organisms contribute to the carbon supply, which supports living organisms. Because synthetic fertilizers either kill or repel bacteria and fungi or don't contain the organic bulk that natural fertilizers do, plants get fed, but soil structure doesn't get built.

A reduction in mycorrhizal fungi means the soils don't get the carbon these fungi deposit via their coating, glomalin. In addition to providing nutrients, mycorrhizal fungi help protect plants. There is chitin, a nitrogen polymer, in the sheaths around roots that is not only a source of nitrogen but keeps parasitic nematodes in check, and some produce antifungal chemicals to ward off competition. By the same token, the antibiotics some microbes produce are missing once these organisms are killed by the use of synthetic fertilizers.

These symbiotic relationships between plants and soil microbes have

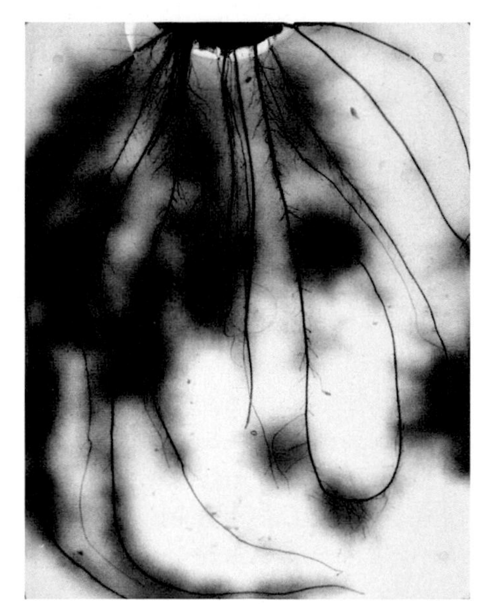

Plant exudates can adjust pH and nutrient supply, and even affect soil quality, by supporting the organisms of the soil food web.

been maintained for some 420 million years. About 60 percent of the carbon produced by plants during photosynthesis is used to synthesize many different exudates that influence pH, the type and diversity of the soil organisms, plant nutrient uptake, plant defense, and soil structure. These carbon-rich root exudates support *Rhizobia* and *Frankia* that fix nitrogen.

It's not just the animals in the soil with which plants have developed relationships. There is one with the soil itself. Plants exchange hydrogen and hydroxyl ions to change the pH so they can get the nutrients they need. In short, plants are capable of modifying the soil chemistry. Some plants, such as grasses, can produce siderophores that help to unlock iron from the soil.

Organic fertilizers are, by definition, full of organic matter, which supports the creation and maintenance of good soils and healthy, diverse soil food webs. These fertilizers contain their nutrients in bulk. Thus, their use also increases the cation exchange capacity (CEC) of soil because it is adding organic matter that holds a charge. Applying organic fertilizer is like constructing a condominium for the organisms that fix nitrogen and mineralize and cycle plant nutrients, ultimately making them available in the very same ionic form as do chemical fertilizers.

Soil structure is important to good gardening. Lose it and conditions become increasingly anaerobic. Plants have a hard time taking up nutrients in such conditions, as we learned in the last chapter. The populations of animals in the soil expel carbon dioxide. Plant roots and beneficial organisms in the rhizosphere around them need oxygen in order to survive. Unless this carbon dioxide is removed, it will react with water and minerals in the soil and create acidic conditions. Some of the carbon dioxide dissolves in water and becomes carbonic acid, which, though a weak acid, still produces hydrogen ions, thus lowering the pH. This also can lead to anaerobic conditions. When this happens, nitrogen fixation diminishes and there is a whole cascade of effects that have negative impacts on plant nutrition.

The addition of organic matter in natural fertilizers helps create a structure in which the carbon dioxide from the microbial activity in the soil is not trapped. This structure also ensures that the proper moisture content, which is critical for the uptake of nutrients, can be much more easily maintained. In fact, if you have sandy soils, using natural fertilizers is the only way to hold in water and may be the only way to really provide adequate nutrients to plants.

Synthetic fertilizers usually contain much higher percentages of nitrogen, phosphorus, and potassium than natural fertilizers, which are derived from plant and animal remains and rock dusts. Thus, the nitrogen content may be as high as 60 percent in some chemical fertilizers, whereas almost all organic fertilizers have about 12 percent nitrogen. However, macro- and micro-arthropods (such as springtails and fungus-feeding mites) and worms, in particular, shun areas where there are synthetic fertilizers. They either don't do well with these high concentrations or their food sources disappear. This further degrades the soil structure and the production of humus. In addition, mycorrhizal fungi and nitrogen-fixing bacteria do not form symbiotic relationships when there are high levels of nitrogen and phosphorus. Mycorrhizal fungi are the largest single source of carbon in soils, and nitrogen-fixing bacteria are a free source of nitrogen, a macronutrient most often in short supply. So, by maintaining a healthy soil food web, gardeners can get carbon, nitrogen, and other nutrients for their plants for free.

When organic fertilizers are applied, the natural substances are broken down into organic compounds by microbes in the soil. Eventually, humus, the end product of composting, is created. This is what good soil is all about, and you can't develop humus without organic matter and microbes. Natural fertilizers are more complete fertilizers. Not only do they provide bulk, which can be used by the soil food web, they also supply many more nutrients at one time, not just nitrogen, phosphorus, or potassium.

Other arguments can be made for the use of natural fertilizers. For one thing, making chemical fertilizers is an extremely energy-intensive activity. These products also have to travel great distances to get to market, which adds to their environmental footprint. More important, many synthetics are anions because farmers need to get their plants fed quickly and need fertilizers that are instantly soluble in water. These anions are easily leached out of the soil and cause serious pollution as they run off into water bodies. Finally, it is often impossible to determine exactly what is in commercial synthetic fertilizers. In the United States, each state has its own labeling laws, and many allow the use of fillers without having to identify their source.

FERTILIZER USE AND POLLUTION

The runoff from thirty-one states and two Canadian provinces enters the Mississippi River and flows to the Gulf of Mexico. When farmers and gardeners in these regions apply excessive amounts of nitrogen and phosphorus fertilizers in soluble form, these nutrients leach into the river. As a result of this pollution, a huge dead zone forms each summer along the coasts of Mississippi, Louisiana, and eastern Texas.

In the spring, warm freshwater runoff from the Mississippi creates a barrier layer atop the Gulf of Mexico. This barrier cuts off the saltier water below from oxygen in the atmosphere. Nitrogen and phosphorus pollution in this freshwater layer cause huge algal blooms to form in the Gulf, shown in red and yellow in the photo. When the algae die, they sink into the saltier water below and are decomposed, which uses up the limited oxygen in that lower layer. This lack of oxygen, a condition known as hypoxia, causes fish and other organisms to avoid the area or to die in massive numbers. The cooler temperatures of winter disrupt the barrier layer, but the cycle starts again each spring.

Although farmers usually get all the blame for the excessive use of synthetic fertilizers, gardeners also play a role. In fact, studies have shown that gardeners use *three times* more synthetic nitrogen per acre than do farmers.

The red and yellow regions are dead zones in the Gulf of Mexico caused by pollution from runoff.

There are arguments against the use of natural fertilizers, too. The first is that you can't feed the world using just organics. For our purposes, however, the argument doesn't really apply. Gardening is a hobby, and it simply does not make sense to use chemicals that degrade the soil and greater natural environment. If you don't consider your garden a hobby, because the food from it is a necessity, then all the more reason to ensure you are not poisoning your family or reducing the ability of your soil to produce.

The strongest argument against the use of natural fertilizers over synthetic ones is that organic fertilizers are not immediately available to plants when they are applied, as are synthetics. This is true. (Although the reverse is also true: synthetics are available too quickly and not for long periods.) Most natural fertilizers are slow-release fertilizers. However, this situation can be overcome in most cases by some careful planning and preseason application of natural fertilizers or by the use of natural fertilizers that are actually readily available after application. This argument only applies at the start of a natural system. Once slow-release natural fertilizers are in the soils for a while, the soil food web releases the nutrients.

NATURAL FERTILIZERS

A fertilizer, by definition, is any soil amendment that can guarantee a certain percentage of nitrogen, phosphorus, and potassium and sometimes other essential nutrients. (In Australia, for example, it is nitrogen, phosphorus, potassium, and sulfur). Laws require that the percentages of these elements be listed on fertilizer packages. This is why every fertilizer package carries the nitrogen–phosphorus–potassium (N–P–K) trilogy on its label. It is also why many people don't consider compost or manures to be fertilizers. These are instead classified as soil amendments because in most cases the amounts of nitrogen, phosphorus, and potassium vary from batch to batch and are usually not measured for labeling purposes.

Most gardeners think the N–P–K trilogy represents the percentage by weight of nitrogen, phosphorus, and potassium in the container. That's close, however, the phosphorus on the label is really the percentage of phosphorus pentoxide (P_2O_5) and the potassium the percentage of potassium oxide (K_2O). This has to do with the way chemists used to measure the phosphorus and potassium. To make the conversions for a pure N–P–K trilogy, you need to multiply the number for phosphorus by 0.44 and that for potassium by 0.83 to determine the actual weights of phosphorus and potassium.

Natural fertilizers are derived from plant and animal by-products as well as rock powders. Although there are as many commercial brands available as there are hosta varieties, they can be grouped into dry fertilizers and liquids. Examples of dry fertilizers are blood, soybean, fish, cottonseed, and alfalfa meals, as well as bat and bird guano and rock phosphate. These are applied either on the soil surface or in the root zone when planting. They can also be mixed into soil when starting a garden. Most are decayed slowly by the soil food web. Liquid fertilizers include things like fish emulsions and kelp extracts. These work faster than most dry fertilizers (bat guano and fish meals are, perhaps, exceptions), provided there is a healthy soil food web to cycle them. Liquids are normally applied to roots and, in some circumstances, sprayed on leaves as a foliar feed. Plants, however, require far more macronutrients than they could ever absorb via leaves, and many nutrients are not mobile once inside plants, so foliar feeding is of limited value.

This dry versus liquid division is good for deciding on an application method. However, now that we know something about the fourteen essential mineral elements, it makes sense to group natural fertilizers by the major nutrient they provide. Some natural fertilizers list more than N–P–K numbers. Those that are used for the other essential nutrients will usually list these and their percentages on the label. You now know what to look for.

It is important to remember that the microbes in the soil decay most natural fertilizers, although some become available due to weathering. The right environment increases the cycling of nutrients by microbes. Warmer temperatures (to a point) will speed up the microbial process if there is adequate moisture. Therefore, the length of time natural nutrients will be available after application varies.

NITROGEN

Alfalfa meal (N–P–K 2–1–3). This is the very same stuff that sustains pet rabbits and horses. Alfalfa meal is a good all-purpose source of nitrogen and contains trace elements as well. Alfalfa meal feeds bacteria and fungi and is usually covered with protozoa, each of which can cycle 10,000 bacteria a day into plant-usable ammonium. It is available for cycling into nutrients by microbes for about 1 to 4 months.

A few caveats are in order. First, because alfalfa is a plant, the meal does contain natural growth hormones, so it can be overused. Use it in the early life of a plant to stimulate growth. The other warning is that alfalfa meal can sometimes contain seeds, so examine it before buying or using it to prevent unwanted weeds. Of course, this can also be seen as an upside: the seed will grow and attract endomycorrhizal fungi, which help to add carbon to your soils.

Bat guano [N-P-K 10-3-1]. Bat guano (feces and urine) is what many of the world's farmers used before synthetic sources of nitrogen were available. It is not only rich in nitrogen, but the nitrogen is in a soluble and readily available form. It can be used at the start of the season or when a quick pick-me-up is needed. It is great for the first-time natural garden when there hasn't been enough time for fertilizers, mulches, and composts to decay and provide sufficient plant nutrients to garden plants. Bat guano is applied in powder form or mixed with water and then applied. It will last from 4 to 6 months.

There are three warnings to consider before using bat guano. If too much is used, it can burn plants, meaning that its presence can pull water out of the cells by osmosis. Next, there are two kinds of bat guano. One is full of nitrogen, and the other is full of phosphorus with much less nitrogen. Read the labels. Finally, bat guano can be expensive, but it lasts for 4 to 6 months. Make sure any guano you buy has been harvested in a sustainable way.

Blood meal [N-P-K 12-0-0]. This meal is made from dried blood collected during the processing of livestock, and it is very high in nitrogen. It doesn't get better, so to speak, because not only is blood meal high in nitrogen, but this nutrient is readily available. Thus, as a fast-acting, natural fertilizer, blood meal that can be used to counter the argument that natural fertilizers are released too slowly. The nutrients in blood meal are released for about 1 to 4 months.

Here are the warnings. With such a high nitrogen content, blood meals can burn plants. It also has an unpleasant odor. Finally, blood meal will attract animals, such as dogs and cats. On the plus side, deer (and moose for those who have them) hate its smell and normally avoid it like the plague. They smell the blood and think a predator has made a kill.

Cottonseed meal (N-P-K 6-0.4-1.5). This is a slow-release, high-nitrogen fertilizer that lasts around 4 months. Cottonseed meal also contains trace elements, such as zinc, copper, manganese, and molybdenum. It is slightly acidic and good for use with acid-soil loving plants, like rhododendrons, azaleas, camellias, and blueberries.

The downside to cottonseed meal is that cotton growers usually use tremendous amounts of pesticides, and these residues can be found in this fertilizer product. Therefore, it is best not to use cottonseed meal on vegetables and fruit crops. There is also the issue of cotton being a genetically modified organism (GMO). This is one crop that may be glyphosate-ready and would not be considered organic.

Corn gluten meal (N-P-K 9-0-0). This is another high-nitrogen natural fertilizer. Corn (maize) gluten is the by-product of the manufacturing of corn syrup. It is effective over a period of 1 to 4 months, depending on rainfall.

The downside, besides expense, is that corn gluten prevents root hairs from developing on germinating seedlings. As you now know, no root hairs, no plants, so you can't use it when you are trying to germinate seeds. Corn gluten meal is much better for use in established lawns and perennial gardens where seeds are not expected to germinate. Corn gluten may come from GMO plants, for those who are concerned about the possibility of ingesting *Bacillus thuringiensis* or who maintain a totally organic garden.

Feather meal (N-P-K 7 to 12-0-0). A by-product of the poultry industry, feather meal is a high-nitrogen plant food. However, it releases the nitrogen much more slowly than bat guano or blood meal. In fact, this is a really slow release natural fertilizer, because feather meal is full of a protein called keratin, which is complex and requires a bit more microbial digestion to fully decay it. Feather meal can continue to be a source of nitrogen for 6+ months.

The use of feather meal may attract dogs, raccoons, and bears.

Fish emulsion (N-P-K 5-2-2). This is heat- and acid-processed fish parts that results in a soluble and highly concentrated nitrogen fertilizer that has a bit more balance of the essential nutrients that the others.

Fish emulsion is full of micronutrients, too, as fish eat lots of things that contain micronutrients. Fish emulsion is soluble and thus quick acting. It needs to be diluted because of its concentrated nutrients. This fertilizer lasts from 1 to 4 months.

These liquids usually smell like rotting fish. Do not (as I once did) apply it to the lawn the same day you want to use it for entertaining. Fish products attract flies, too, which help in the decay process. They can attract bears (at least here in Alaska), and some dogs love to roll in grass after it has been applied. Things usually return to normal within a few days.

Fish meal [N-P-K 10-6-2]. This meal is ground fish parts that are heated and dried. In addition to nitrogen, fish meals are a good source of phosphorus. Like the other fish-based fertilizers, fish meal lasts from 1 to 4 months. Fish meal is not readily soluble and not as quick starting as other fish fertilizers, but it is a decent phosphorus source, which the others are not.

Fish meals smell for a few days, although they do not smell as strongly as fish emulsions. They also can attract flies, bears, and dogs. Fish meals are heat processed, which results in a loss of some of the enzymes and other proteins, vitamins, and even micronutrients that remain available in the other fish products.

Fish powder [N-P-K 12-0.25-1]. Fish powder is another heat-processed fish material, only this has a highly soluble form of nitrogen. Because fish powder is water-soluble, it can leach out of soils easily. It is fast acting and used up within 1 month under normal conditions. Like blood meal, fish powder is another fertilizer that acts almost as quickly as synthetic fertilizers.

Hydrolyzed fish [N-P-K 4-2-2]. This product is fish that have been enzymatically digested in large tanks instead of being heat processed. The remaining liquid retains many more proteins and other compounds than if heated. This medium-acting natural fertilizer lasts up to 5 months.

The downside is that hydrolyzed fish can be relatively expensive.

Human hair [N-P-K 18-0-0]. Wow, talk about a lot of nitrogen! For every 7 pounds of hair, there is around 1 pound of nitrogen. Here is a natural fertilizer that exceeds the normal percentage for organic fertilizers, with a nitrogen content closer to a chemical lawn food. You won't find this bagged and on the market, but you can get it for free. Hair, fortunately for those of you who still have yours, is not readily soluble and does not readily decay, so the high nitrogen content doesn't harm the soil food web. In fact hair is a very slow-release source of nitrogen, lasting from 1 to 2 years.

Human urine (N–P–K 15–2–2). Urine is another fertilizer that has a relatively high nitrogen content for natural fertilizers. It gets my vote for the first fertilizer ever used. It is sterile when first produced and contains urea, which when pure has an N–P–K value of 46–0–0. Soil bacteria quickly convert urea into ammonium. If the soil pH is acidic, this is the form mostly taken up by plants. If the pH is basic, which encourages nitrifying bacteria populations, the ammonium is converted into nitrates for uptake. Either way, urine can burn tender plants (and microbes) if not diluted with at least 8 parts water to 1 part urine. (However, from 60+ years of personal experience, I can tell you there is no problem with occasional application to trees and shrubs, especially when no one is looking.) Used undiluted, it is great for application to compost piles. One application will last for 1 or 2 weeks.

The downside of human urine is concerns about the possible presence of heavy metals, such as mercury, as well as antibiotics. Do not use urine on food crops.

Soybean meal (N–P–K 7–2–1). Soybean meal is one of my favorite all-purpose fertilizers because it supplies lots of carbon as well as nitrogen, doesn't smell, stores well, and works in a spreader or when broadcast by hand. It quickly attracts both fungi and bacteria, which slowly release its nutrients for 3 or 4 months.

On the downside, soybean is one of the biggest GMO crops. The insertion of bacterial genes into a plant genome results in a protein molecule being produced that has never been in that plant or in food derived from it. These may require special microbes to break down, both in the soil and in animals. Some argue that this is the cause of an increase in soy allergies and autoimmune and digestive problems. Studies on cattle, poultry, and swine indicate that greater caution should be used with GMO crops. Now that you know how plants eat, you have the tools to follow the debate on this issue, which includes deciding whether to use GMO-containing fertilizers.

Chilean nitrate (N–P–K 16–0–0). This is a quick-acting, fully soluble mineral fertilizer that has percentages of nitrogen over the norm for natural fertilizers. It can be used to boost nitrogen without increasing anything else. While not usually allowed under certified organic systems because it is not technically organic, Chilean nitrate is a natural nitrate mined in the desert of northern Chile. Some organic certifying agencies allow its use in limited quantities, particularly where soils remain cool so microbial mineralization of nitrogen is low and slow and a quick release is needed. Natural nitrate soda is fast acting because it is soluble immediately and does not require microbial activity to convert it into ions. It is mobile in the soil and will only be readily available for 1 or 2 weeks or until the next good rain.

Natural nitrate soda does contain 26 percent sodium, which is fine in small quantities. This fertilizer should not be used where there is already high sodium content in soils, such as in deserts and dry regions. Sodium can also react with clay, which makes it difficult to build soil structure and aeration. This is one natural fertilizer whose use does not improve soil tilth (structure), so Chilean nitrate should not be the sole source of a garden's nitrogen. It is best to limit its use to start a new garden and at the beginning of the season, especially in gardens where it takes a long time to warm up.

PHOSPHORUS

Animal bone meal (N-P-K 3-15-0). This fertilizer is made by steam processing and grinding bones. The phosphorus in bone meal is very readily available and lasts 1 to 4 months.

Unfortunately, bone meal does not work well as a source of phosphorus unless the pH is below 7. You have to read labels. Some brands list the percentage of phosphorus and others phosphate, which is the norm. It also may attract animals until the attractant odor disappears 1 to 2 weeks after application.

Bat guano (N-P-K 3-10-1). Guano (the feces and urine of cave-dwelling bats) comes in a high-nitrogen form and one with lots of phosphorus. Because phosphorus is quickly tied up in soil, high-phosphorus bat guano will last 1 to 4 months. The best releasers of this nutrient are mycorrhizal fungi. Too much phosphorus, however, limits their presence, so go easy and make sure your laboratory knows you rely on mycorrhizal fungi when they develop recommendations based on your soil tests.

Unless you are starting a garden and have not had time to prepare it a few months ahead of planting, use this fertilizer carefully and only when a testing laboratory suggests guano as a source of phosphorus.

Colloidal rock phosphate (N-P-K 0-2.5-0). Sometimes called soft rock phosphate, this material consists of clay particles surrounded by phosphate. The clay helps to improve the soil's cation and anion exchange capacities. This stuff lasts years, slowly releasing phosphorus that makes its way to the roots via mycorrhizal fungi or by diffusion. Some gardeners claim that because of the way phosphorus is taken up into plants, colloidal rock phosphate actually works better the second and third years. It remains available for 3 to 5 years, depending on rainfall and watering.

Placement is key with rock phosphates. They work best when placed where roots will intercept the particles.

Crab shell meal (N-P-K 2-3-0). This is a good source of phosphorus, if you can find it. In addition to the nitrogen and phosphorus content, crab shell meal also contains lots of calcium and trace elements. It also contains chitin, making it a great fungal food. Chitin can also control high populations of nematodes.

Nests of the Peruvian booby are made of guano.

POTASSIUM

Greensand [N-P-K 0-0-7]. Often called New Jersey Greensand or glauconite, this mineral was formed on ancient sea beds and contains more than thirty elements, including calcium, magnesium, and iron and other micronutrients, as well as potassium. It has to be weathered and is slow to release nutrients, so it lasts 2 or 3 years.

Wood ashes [N-P-K 0-1-3]. Wood ashes or pot ash, as they were referred to in the old days, have been in use for centuries as a source of potassium. They are free for those with a fireplace, but care must be taken not to use ash from treated charcoals or woods, as harmful chemical residues may remain. Ashes are alkaline and will increase pH, so they should not be used in soils that are already alkaline. They last 1 to 2 months.

Sulfate of potash [N-P-K 0-0-2]. This mineral salt is often labeled Sul-Po-Mag. It is very soluble in water and contains 23 percent sulfur, 22 percent potash, and 11 percent magnesium. Some gardeners do not consider it to be organic, because it does not add bulk to the soils. The solubility of sulfate of potash makes it very quick acting, and it lasts only about 2 months.

CALCIUM

Calcitic limestone [N-P-K 0-0-0]. This material is from sedimentary rock that contains calcium carbonate ($CaCO_3$). A slab of limestone will last thousands of years, but when powdered and exposed to water and carbon dioxide, which form an acid, it breaks down. How long the calcium ions remain in the soil has much to do with its pH and organic matter and clay contents. It is more important to know that you can only raise pH a point or so per growing season. The best time to apply is in the autumn for the following growing season.

Dolomitic limestone [N-P-K 0-0-0]. In addition to providing calcium, dolomite or dolomitic limestone ($CaMg[CO_3]_2$) is added to soils as a magnesium source because it contains about 10 percent of this element. It tends to bind soils and too much magnesium impacts the uptake of other nutrients, so make sure your soils test shows you actually need more magnesium. Again, the efficacy and duration of limestone in the soil is related to the existing pH, water, and carbon dioxide, another great example of why testing is so important.

MICRONUTRIENTS

Shrimp shell meal (N-P-K 5-8-15). This meal is made by grinding up the heads and shells of shrimp. Shrimp shell meal also contains lots of trace elements, as you would expect from a creature that lives in the ocean. In addition to phosphorus and potassium, it also contains about 15 percent calcium and about 20 percent chitin. Shrimp shell meal is a slow-release fertilizer that lasts 3 to 6 months.

Kelp meal (N-P-K 0-0-0 to 1-0-4). Kelp meal is made from kelp, a type of seaweed. Seaweeds contain up to sixty elements, including all the trace minerals that plants need. Kelp meal also contains natural plant growth hormones, too. The nutrients are generally available about 1 or 2 months after application, and they are then slowly released during the 4 or 5 months that it takes the meal to decay.

The negative is that, while often very effective for hydroponics, many of these nutrients are already in the soil. Although trace minerals are essential, they are needed in very minute quantities. In addition, it is important to make sure the kelp is harvested sustainably.

Kelp powder (N-P-K 0-0-0 to 1-0-4). Kelp meal is further ground to make kelp powder. This makes it more soluble and thus more readily available to microbes, which quickly make the nutrients available to plants. It lasts for up to 1 month.

Kelp on the beach of Chowiet Island, Alaska. Kelp is full of micronutrients.

Liquid kelp (N–P–K 0–0–0 to 1–0–4). This is another good source of micronutrients that are instantly available to plants when applied. Most brands of liquid kelp are made by enzymatically digesting kelp, which preserves more of the plant's growth hormones than the other processes used to convert kelp. It lasts 1 to 3 weeks in the soil.

BIOFERTILIZERS

More and more gardeners are using live agents to help produce plant foods. Again, while not fitting the current legal definitions of fertilizers because it is difficult or impossible to quantify their N–P–K contents, there are plenty of microorganisms that produce or are intimately involved in the production of plant nutrients. These go beyond the decaying and cycling microbes. Microbes described as biofertilizers produce nutrients.

***Rhizobia* and *Frankia* (for nitrogen).** Both of these bacteria are symbiotic nitrogen fixers. They can produce a lot of useable nitrogen. The use of *Rhizobia* to grow legumes (such as soybeans, locust trees, and wisteria) and to produce nitrogen for soils is a well-established practice. *Frankia* were discovered much later and their use is still developing. Remember, there is incredible specificity between plant and microbe when it comes to nitrogen-fixing diazotrophs. This has to be taken into account when trying to employ them in soil.

***Azotobacter* and *Azospirillum* (for nitrogen).** These two bacteria

These tiny round root nodules are filled with specialized bacteria that capture nitrogen from the air and trade it with the legume for sugars the plant produces. The pinkish color comes from a compound similar to the hemoglobin in red blood cells.

genera are free-living and produce nitrogen without entering into a symbiotic relationship with plants. *Azotobacter* and *Azospirillum* are frequently used when growing cereal crops. Researchers are working to develop mixtures for gardening use.

Phosphate-solubilizing bacteria and fungi (for phosphorus). Certain types of nonsymbiotic bacteria (*Bacillus megaterium* var. *phosphaticum, Bacillus subtilis, Bacillus circulans, Pseudomonas striata*) can free up insoluble phosphorus and either make it directly available to plants or put it into the diffusion stream that brings phosphorus to plants. They are called phosphobacterins.

Similarly, certain fungi (*Penicillium* species, *Aspergillus awamori*) also free up phosphorus. They act indirectly on the insoluble phosphorus by producing organic acids as they go about their business. These acids break the bonds that tie up phosphorus in the soil.

Mycorrhizal fungi (for phosphorus, copper, zinc, molybdenum, and nitrogen). These fungi form mycorrhizae with plant roots. In return for the plant exudates that supply carbon (which fungi cannot make), mycorrhizal fungi obtain phosphorus from the soil and make it available to plants. Their long hyphae extend root surfaces up to hundreds of times, so interception for this and other nutrients is high. Mycorrhizae are the norm in the natural world, with over 90 percent of plants entering into such a relationship.

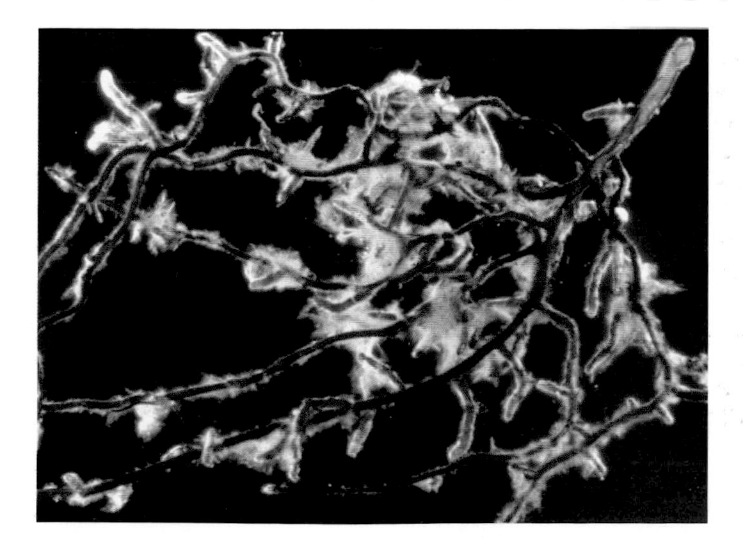

Mycorrhizal fungi may provide as much as 80 percent of a plant's phosphorus and 60 percent of its copper along with other essential nutrients. This is *Suillus* infecting a pine root.

Some experiments have suggested that these fungi can deliver 80 percent of a plant's phosphorus, 60 percent of its copper, 25 percent of its nitrogen, 25 percent of its zinc, and 10 percent of its potassium. This is free—not only economically, but work-free. Mycorrhizal fungi are the reason trees do so well with so little care.

Plant growth promoting rhizobacteria (PGPR). Rhizobacteria are root-colonizing bacteria that form symbiotic relationships with plants. The most well-known species is *Pseudomonas fluorescens*, but PGPR are actually a whole host of bacteria that aid in the synthesis of nutrients, positively influencing root growth and thus plant nutrition. The major function studied so far is the way these bacteria help mycorrhizal fungi with obtaining phosphorus, perhaps helping the fungi break bonds. Some PGPR produce alkaline phosphatase, an enzyme that breaks down phosphate bonds.

Compost (N-P-K varies, but low). Compost is technically not a fertilizer, but rather a soil amendment. However, compost is full of microbes and humus, and it provides a great environment for the microbial activity needed to cycle or decay natural materials into nutrients. Compost does contain nutrients because organic matter and any clay in the compost will have an impact on its CEC. Many would argue it should be tested for nutrients along with microbial content so that the gardener can have an understanding of what is being applied and how its application might impact soil laboratory recommendations.

Earthworm castings concentrate nutrients, which is why they are so popular with organic gardeners. They contain 10 times more potassium, 5 times more nitrogen, 7 times more phosphorus, 3 times more magnesium, 1.5 times more calcium, and 1.4 times more humus than the soil that went into the worm.

A big caveat is that compost must actually be composted. Anything that has not gone through the complete process, and many home compost piles never do, is not compost. Partially composted material can create problems, including tying up nitrogen, as the composting process is completed next to the plant instead of in a pile.

Earthworm castings (N–P–K varies, but high). Worm castings, another "almost fertilizer," but this time because it's next to impossible to supply a guaranteed N–P–K analysis, also contain a lot of microbes. The worms ingest organic matter, but what they are really after are bacteria, protozoa, and fungi, some of which they digest. They process the rest of the material into castings with a higher concentration of organic matter and great N–P–K, as well as calcium, copper, zinc, and other minerals, than in the source material.

Manures (N–P–K varies). Technically not fertilizers, livestock manures in the modern age have to be carefully assessed to determine what they contain in the way of antibiotics, hormones, and medicines. These often do not break down during the composting process. The use of manures can lead to the buildup of salts and heavy metals. In addition, if they are not composted properly, manures can be the source of *E. coli*. In the home garden, manures must be completely composted for at least 72 hours at a minimum of 131°F (55°C) before use.

MAXIMIZE THE AVAILABILITY OF NUTRIENTS

Lots of things affect the availability of nutrients from organic fertilizers, and all of them relate to the ability of the minerals to be decomposed to their ionic forms. This decomposition is usually a direct result of the activity of organisms at the base of the soil food web: bacteria, Archaea, and fungi.

The temperature of the soil can make a difference in availability because the microbes that release nutrients are generally more active at warmer temperatures. In addition, you should ensure that your soil's pH is in the right zone to maximize the numbers of unencumbered ions so that as few as possible are rendered unavailable to plants. However, the number one rule for maximizing the availability of fertilizers is to know what nutrients are already in your soil and as much as possible about its characteristics. Have a soil test done before you set up a feeding program. Retest the soil at the end of the season or the start of the next (or as otherwise advised by your laboratory) to see how things are trending. Thereafter, test garden soils every 2 or 3 years and adjust your program accordingly.

MAKE YOUR OWN FERTILIZER

Now that you know something about how a plant eats, the next step might be to make your own natural fertilizers. Sure, you can buy all you need in commercial mixes. But how difficult can it be with only fourteen mineral elements needed? Besides, if a plant can take these and make millions of different kinds of molecular compounds, we should be able to make a few simple recipes ourselves.

There are several ways to make your own plant foods. Each is dependent on what kind of gardens you have and what kind of gardener you are, as well as the sources of supplies. Some gardeners make one fertilizer that can be used with most plants. Others develop specific mixes for different kinds of plants, such as lawn grasses, vegetables, annual flowers, and trees and shrubs. Still others make mixtures for different times of the year. You can even make mixes of fast-acting, soluble natural fertilizers to use on unprepared soils or when transplanting, to ensure that plants start growing with sufficient nutrients on hand. In fact, some natural fertilizing systems rely on a starter formula for use early each season, followed later in the season by one or more slow-acting mixes.

The ingredients in the following recipes are combined by volume, not by weight. So, in addition to a bucket to keep it in, you'll need a large scoop or cup to use for measuring. It doesn't matter what size the scoop is, as long as you use it for all of the ingredients in the mix.

► **All-Purpose Vegetable and Annual Organic Fertilizers** There are many recipes for making a natural fertilizer mix for general use. These need to include nitrogen, phosphorus, potassium, and trace elements likely missing from the soil, as these are most likely the limiting nutrients.

Both of the following recipes should be adjusted in accordance with your soil test results. If the soil doesn't need as much of a nutrient, you can leave some of the ingredient containing that nutrient out. In short, adjust the recipe in keeping with the testing laboratory's recommendations.

In my experience, adding 0.5 inch (1.3 cm) of compost, or 6 inches (15 cm) of grass clippings when there is no compost, as a cover to these fertilizer mixes increases their efficiency in helping with microbial activity. The use of mulches always makes sense, as there are no bare soils in nature.

Grandpa Al's Can't Fail Recipe

4 parts fish meal or any of the nitrogen-supplying meals such as soy or cottonseed meal
1 part kelp meal
1 part rock phosphate or ¾ part bone meal
1 part dolomitic limestone, or 1 part calcitic limestone, or ⅓ part dolomitic limestone, ⅓ part Sul-Po-Mag, and ⅓ part calcitic limestone

This mix is applied at a rate of 1 to 2 gallons per 100 square feet (3.8 to 7.6 liters per 9.3 m^2) at the start of the season (that is, prior to planting), banded into the root zone. Side dress under the mulch every 4 weeks.

Nitrogen is provided by the meals. Phosphorus is provided by the rock phosphate or bone meal. Potassium is provided by the sulfate of potash and the kelp, which also supplies micronutrients along with cofactor minerals that help in enzymatic reactions.

Note that the use of limestone here is not to change the pH of the soil. It is to counter the acidic pH created by the other ingredients in the mixture. If your soils are very alkaline (and you know this because of soil tests), some lime is O.K. and needed, but you should reduce the amount of dolomitic limestone added because magnesium will tie up other nutrients at high pH.

Steve Solomon's Recipe

3 parts cottonseed meal
1 part blood meal
1 part dolomitic or calcitic limestone
½ part bone meal
½ part kelp meal

Apply this at a rate of 6 quarts per 100 square feet (5.7 liters per 9.3 m^2). Band it into the root zone for quicker results, but only if you really need them. A system that has already been natural for a few years won't need mixing in. Side dress after removing mulch once every 3 or 4 weeks; reapply the mulch after application.

The nitrogen is supplied by the cottonseed meal and the blood meal, which supplies a much more immediate pool of nitrogen. Phosphorus comes from the bone meal, and potassium is provided by the kelp meal. As with the previous recipe, the use of limestone is not to adjust the soil pH.

► **All-Purpose Vegetable and Annual Starter Solutions** Again, the big negative to some gardeners in using natural fertilizers is their slow startup time. Some natural fertilizers are more soluble than others, however. This is one reason some gardeners use a liquid solution, rather than meals or powdered mixes, because nutrients that are in the water solution are immediately available.

Quick Starter for Planting

1½ parts fish emulsion or fish powder, or ½ part bat guano
1 part liquid kelp

Dilute with water, according to the label instructions for the fish emulsion, fish powder, or bat guano. These ingredients provide a very quickly available source of nitrogen, phosphorus, and potassium. The kelp provides the trace minerals. The trick is to make sure this source of soluble nutrients gets to the root zone quickly. You can soak transplants in a diluted solution of 1 part starter to 4 parts water for a while before transplanting.

This is the only mixture that comes with warnings. First, do not use this mix in clay soils as it will deflocculate, or break apart, the clay. Second, this mixture has a lot of soluble nitrogen in it and must be mixed at the right dilution rate and then applied at the right application rate. The starting dilution rates to use are 3 tablespoons of bat guano per gallon (3.8 liters) of water, 1 teaspoon of fish powder per gallon of water, or 2 teaspoons of fish emulsion per gallon of water. Application rates are often included on the label. If not, assume that a gallon (3.8 liters) will service the plants in a 4 to 5 square foot (0.4 to 0.5 m^2) garden. In fact, it is always a good idea to test the formula on sample seedlings. (Finally, you have something to do with the extra seeds in a packet.) If they don't die after application, you have an acceptable formulation and will see them respond quickly.

► **Lawn Fertilizers** Lawns are a classic sink for commercial fertilizers because most chemical gardeners violate the Law of Return by removing grass clippings and raking fallen leaves instead of mulching them in. Once a gardener stops removing leaves and grass clippings, lawns need very little fertilizer input. This is especially true if clover moves in, as it hosts *Rhizobia* bacteria that fix nitrogen.

Some want a greener lawn. You can try spreading a bit of compost, 0.5 inch (1.3 cm) deep, or use Wayne Lewis's Graceland Lawn Food, a good general-purpose lawn food tested extensively by my colleague Wayne Lewis.

Wayne Lewis's Graceland Lawn Food

1 part soybean meal or chicken litter meal

1 part granulated molasses

A 50-pound (23-kg) bag of each, mixed, will fertilize 2000 square feet (186 m^2) of lawn. Apply it after weekly watering alone no longer keeps the lawn green. Unlike commercial lawn fertilizers, you really cannot apply too much and it surely will not burn the lawn. If you are just starting out or are unsure, apply it once in the spring and see how you feel after the first month. It helps to leave grass clippings, which will provide nitrogen as per the Law of Return as well as foster microbes that will cycle the soybean and the molasses more quickly.

The soybean meal provides nitrogen and substrates for fungi, and the granulated molasses is a terrific microbial stimulant as well as a fungal food. The combination results in lawns getting ample nitrogen. If you really need to load on the nitrogen (based on soil test results), add 1 part nitrate of soda or bone meal to the recipe. These won't last as long as the chicken litter meal or soybean meal, but you will see some quicker reactions. You'll also have to mow more often.

COMMERCIAL NATURAL FERTILIZERS

These days, there are as many organic mixes on the market as there are chemical ones. Some are single-nutrient fertilizers, and others are more complete mixtures. As fertilizers, each has to comply with labeling laws and will display the N–P–K trilogy, at a minimum. Natural fertilizers add organic bulk to soils, so everything is of benefit. Chemical fertilizers don't contain organic matter but rather fillers, which may or may not be of benefit—but are never as beneficial as organic matter.

It may be difficult to get a perfect fit between what your soil needs and what is available in a commercial mix on the shelf. Because nitrogen is the nutrient most used by plants and the one that is most often inadequate in supply (Do I hear Von Liebig's law?), soil testing laboratories usually suggest the gardener apply an N–P–K formula that first meets the nitrogen needs of the soil.

APPLYING FERTILIZER

There are lots of different ways to apply fertilizer, and each should be explored. The most prevalent is broadcasting, a technique used to spread fertilizers over large areas. Fertilizer in granular or powder form is spread over the surface by hand or by using a drop spreader. A hand-cranked broadcast spreader works on a lawn. Although fine for lawns or a large garden, broadcasting is not very efficient in some situations. The areas between rows and plants receive fertilizer, which is a waste and encourages weeds. Broadcasting also places phosphorus on the surface, where it pretty much remains unavailable to plants.

Banding, on the other hand, is a way to apply fertilizer (and compost and humus) to the soil before planting so plants will get off to a quick start. It involves putting a band of fertilizers in the root zone. This band should be about 2 inches (5 cm) away from where the plants or seeds will go in and be about 2 inches (5 cm) deeper than where seeds are planted. For transplants, place the band 3 to 4 inches (7.5 to 10 cm) below where transplants will go into the garden.

Banding is used especially when applying nutrients that are immobile in soil, particularly phosphorus and potassium, which need to be near the roots for efficient uptake. Studies have shown a 50 percent increase in phosphorus and potassium uptake when fertilizer is placed in a band, rather than broadcast.

You can spot band by putting a bit of fertilizer in a hole and planting around it. This is a great way to apply immobile phosphates and potassium in autumn, as well as some of fertilizers that take a bit of time before they are mineralized, such as colloidal phosphate and feather meal. It is also a great system for starting plants.

Garden books often tout side dressing for the second and third applications during a growing season. This entails placing the fertilizer along the row or beside a plant, usually at midseason. Place the fertilizer under mulch, or you can bury the fertilizer a bit or insert it into the top 1 or 2 inches (2.5 or 5 cm) of soil by making a slit in the mulch and soil along the side of a row or an individual plant with a knife or trowel and pouring the fertilizer in. Some gardeners use a dilute mix in water and pour on this mix as a weekly side dressing, but this probably isn't the most efficient way to apply it due to runoff and drainage. Also, the sudden and relatively frequent applications change all sorts of conditions that affect how nutrients move to and into roots. Besides, it seems like a lot of work to make so many applications.

I've already mentioned what I think are the limitations of foliar spraying fertilizers. This practice is fine for a temporary quick fix for deficiencies of iron and zinc, which are mobile. The macronutrients can

Banding organic fertilizers is a good way to put them in the root zone, where they are needed.

sometimes enter the leaves this way, too, but gardeners cannot supply plants with enough of these by foliar spraying to keep the plants thriving. The bottom line is you should not rely on foliar feeding for anything but the most mobile nutrients.

Once established, most perennials, trees, and shrubs don't need much fertilizer if they are maintained in a natural system and the Law of Return is allowed to operate. The exceptions might be at planting time to ensure the plant gets nutrients while it establishes a permanent root system. If a plant is pruned often or has lots of fruit removed for consumption instead of decaying on the ground and returning nutrients to the soil, the application of fertilizer might be in order. Again, only a test will really tell. If you do apply fertilizer to trees and shrubs, apply it on or near the soil surface. In most instances, feeder roots for trees, in particular, are within the top few inches.

More and more biological organisms are being marketed to help feed plants. For example, *Rhizobia* inoculates for legumes are common. However, because these plant-microbe relationships are very specific, you should do some research to ensure a match before wasting time and money buying them. Rather than purchasing these bacteria, you can gather some root material and open up the nodules that are created. The pink stuff is coated with *Rhizobia*. Roll seeds in it or mix it into soil.

Mycorrhizal fungi mixes are also available commercially. Fungal spores and propagules only germinate when they receive root exudates, so contact with roots helps. Similarly, rolling legume seed or roots into *Rhizobia* mix will help infect the legume plant when the seed germinates.

Many mycorrhizal fungi are ubiquitous in established natural systems. Therefore, they usually only need to be introduced in new systems, when annuals and vegetables are started indoors, and when plants are grown in compost, which doesn't have much by way of mycorrhizal spores. There is the question of specificity of fungi to host plant, but mycorrhizal mixes usually contain enough different species to infect some roots. It's not quite as easy with *Rhizobia* because of a higher specificity, but if you have a legume such as clover or soybeans or peanuts, you can cut open nodules and use the contents on similar legumes.

You can also go out and collect local mycorrhizal fungi and spores. You can't see the endomycorrhizal fungi without a microscope and staining them, and it takes a trained eye and a good magnifying glass to find ectomycorrhizal fungi (unless they are fruiting, in which case their mushrooms are a great source of spores). However, by taking soil from root areas of thriving plants you will surely get spores, and you can inoculate the soil of the new plants by adding some of this soil to their mix near or on their roots

SUMMING UP

Soil testing must become part of the thinking gardener's activities. We have the knowledge to follow how nutrients move to plants and get into them and a rudimentary understanding of what they do inside. However, we can't see them in the soil or in the plant. Only by testing soils can the gardener know what needs to be adjusted to ensure the presence of all the essential nutrients in adequate quantities.

You have the information to make educated choices, rather than just blindly following some label on a box because it makes you (and some ad agency) feel good. Now that you know about the botany, cellular biology, soil biology, and chemistry involved in how plants eat, consider yourself their gourmet chef. No more generic fast foods that ruin the soil food web, change the pH, introduce competing nutrients, or chemically tie them up. After a soil test, of course, and based on the resulting recommendations, you can prepare meals that are perfectly suited to your plants.

- **Use natural fertilizers** because they feed the soil food web that creates soil structure and cycles nutrients to plants.
- **Natural fertilizers** are derived from plant and animal by-products as well as rock.
- **Nutrients** in all but the most soluble natural fertilizers are only available when they are mineralized by the soil food web.

- **Good natural sources** of nitrogen include alfalfa meal, bat guano, blood meal, cottonseed meal, corn gluten meal, feather meal, fish emulsion, fish meal, fish powder, hydrolyzed fish, human hair, human urine, soybean meal, and Chilean nitrate.
- **Phosphorus** is provided by animal bone meal, bat guano, colloidal rock phosphates, and crab shell meal.
- **Greensand,** wood ashes, and sulfate of potash are good sources of potassium.
- **Calcitic limestone** and dolomitic limestone are commonly applied for calcium deficiencies and to adjust the soil pH.
- **Good natural sources** of trace elements include shrimp shell meal, kelp meal, kelp powder, and liquid kelp.
- **Biofertilizers** include symbiotic, nitrogen-fixing *Rhizobia* and *Frankia;* free-living, nitrogen-fixing *Azotobacter* and *Azospirillum;* and phosphate-solubilizing bacteria. Some fungi increase the movement of phosphorus to roots, and mycorrhizal fungi deliver phosphorus, copper, zinc, molybdenum, and nitrogen to plant roots.
- **Growth-promoting** rhizobacteria form symbiotic relationships with plants that aid in the synthesis of nitrogen.
- **Earthworm castings** concentrate nutrients and create humus.
- **You can make** your own fertilizer mixes with the recipes included in this chapter.
- **The nutrient amounts** in commercial natural fertilizers may not meet your soil's exact needs. In that case, buy according to your soil's nitrogen needs.

- **Apply fertilizer** by broadcasting, banding below the root zone, or side dressing, depending on needs.
- **Banding** is especially useful for phosphorus and potassium.
- **Mycorrhizal fungi** and *Rhizobia* can be purchased or collected locally and applied to the roots.

Epilogue

THE MORE I STUDY plant cells and their operation, the more I start to examine our own world and indeed our universe in terms of cellular makeup and function. When you think about it, the world could be just one big cell. All the functions I studied can be analogized to something, just as microtubules are like railroads. Our world could be just an organelle in one cell in the universe. It is truly a mind-boggling comparison, kind of like the worlds in Dr. Seuss's *Horton Hears a Who*.

Think about it. Everything in our world can be likened to something in a cell. Membranes? We have them: border crossing guards that regulate traffic between countries. Tubules and microfibrils? We have them. They are the tracks and the roads along which we travel. Those complicated route maps at the back of airline magazines look like some of the diagrams I have seen that show how proteins move through a cell. Enzymes? Hormones? Metabolites? Our world has analogous entities, along with garbage collections sites, power-generating factories, and package assembly and delivery systems that are exactly like those in a tiny plant cell. We have signaling systems, water transportation systems, and the list goes on.

By the time I finished the research for this book, I had developed my own cell theory of life: we are all part of a cell, which is part of a larger universe of cells. I will forever look at a power generator and think of mitochondria. I will travel along roads, bike paths, railroads, and air routes around the country and think of tubules and microfibrils. I might even start to think of myself as a cell-made enzyme, strung together in a ribosome to become an agent dedicated to speeding up the return to organics and the soil food web by gardeners and farmers. I will see ribosome factories, endoplasmic reticulum manufacturing centers, and Golgi body packaging, assembly, and delivery sites.

Could it be that we and all the other organisms on Earth are all parts of one big cell? Part of a system as bacteria are inside humans, carrying out functions so the larger organism will thrive? Is our cell and the universe it's in small enough so that we could be part of a plant and not have any idea?

It is so easy to find analogies to cells, cell parts, and even the chemical reactions that occur in cells. Everything we do affects that system. How we use water and nutrients is similar, right on down to how we are recycled in the end. Or is that the beginning? Maybe our whole universe is just one cell, connected to another universe by some plasmodesmata in a sea of similar cells. It is amazing to contemplate.

At the same time, I cannot stop thinking about how all of the cellular activity that is life started. It's not just the 15 trillion cells in the tree outside your window, but all the cells of plants everywhere. They operate the same way. How did they come to be?

I can imagine a bunch of membranes forming, like plastic shopping bags, floating around and occasionally entrapping some flotsam or jetsam in the primordial soup. Things would eventually accumulate inside them to the point that chemical reactions start to occur. Maybe some waste vacuole had the right mix of ingredients to do something with less energy, thus becoming a better vacuole. A protozoan swam into the new vacuole and found it could live there if it just provided some energy to its host. The protozoan divides and thrives, and when the vacuole divides, each part gets protozoa.

Just thinking about the process that duplicates DNA and RNA leaves

A CELL OR THE UNIVERSE?

Plant cells are distinct from animal cells in having an additional external envelope, the cell wall, outside the plasma membrane. In this micrograph, the cell wall, made of proteins and polysaccharides, appears as the thin layer between the cells. The wall defines the shape of the cell and expands as it grows.

Nutrients can travel through and along cell walls in the apoplastic pathway. Eventually they must either exit via the same path or enter the cell by traveling through the plasmalemma, the thin, double-layered phospholipid membrane that is adjacent to the cell wall. Once inside the cell, nutrients mix with the cytoplasm and are converted to whatever building blocks the cell needs by the myriad enzymes floating in it. These new compounds are then assimilated or transported to other cells.

The prominent round organelle in the cell is the nucleus, which contains a smaller nucleolus (red). The double-layered membrane around the nucleus is extremely important for controlling what gets in and out of the nuclear envelope. It is here that the DNA is stored. Some nutrients are stored in vacuoles, seen as the yellow areas in the cytoplasm. These nutrients are available for transport to where they are needed.

Despite knowing what goes on in a plant cell, one can't help but see a similarity between a cell and the universe: millions of tiny parts, each affecting the others, floating in a cytoplasmic soup. Is our universe nothing more than just one cell, part of larger collection of cells? Yes, it helps to know what goes on in a cell. However, we should appreciate the mystery, too.

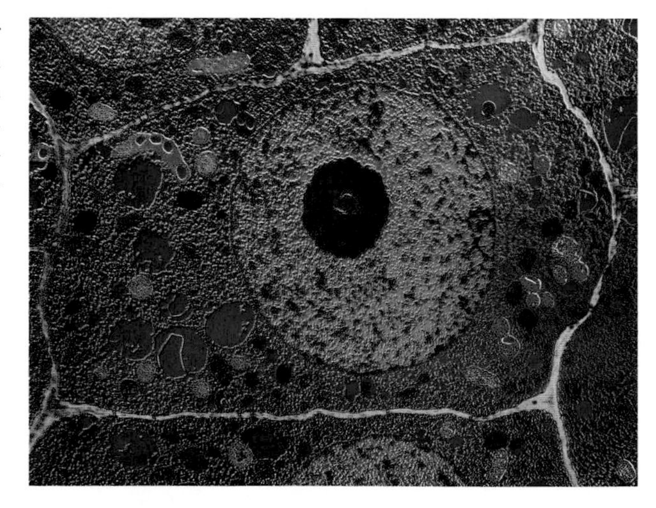

This false-color transmission electron micrograph shows a cell in the root tip of a maize plant.

my head spinning. And the little machines, the spindles that lay down the cellulose in every plant cell wall like spinning cotton candy. Whew. Duplication of cells and cell parts, an enzyme for every one of the mind-boggling number of reactions, a transport protein for each nutrient to move across membranes, and membranes whose molecules line up like Alaskan yaks to keep young ones in and evil ones out. Amazing, simply amazing—and something we gardeners give very little thought to. As little thought as we do to the wonder that a plant can take seventeen nutrients, grow, and recreate itself.

I am learning to look at a picture of a cell and see myself reduced down to the size of a cell-traveling nutrient. When I work in the gardens these days, I think about how my plants are eating, synthesizing the essential elements into everything they need, and, come to think of it, I need as well. But I also now look at a picture of a cell and see a whole universe. When I'm working with my plants, I see myself as just a mere enzyme performing my assigned cellular functions. What goes on in a plant cell is all too amazing to be ignored.

Resources

The reader is advised to use an Internet web search engine to learn more about the organelles, nutrient elements, and cellular processes. The Internet is replete with information on all the subjects in this book. One place to concentrate is on YouTube, which has numerous lectures geared toward almost all levels of interest and background. In addition, many universities are placing all of their classes on the Internet as video or podcast lectures. These are not listed here, but can often be found by using a search engine's Video section or by searching the catalog of any of these online university offerings.

Carbohydrates, Lipids, Nucleotides, and Proteins

http://www.stolaf.edu/people/giannini/flashanimat/carbohydrates/glucose.swf
http://www.hcs.ohio-state.edu/hcs300/biochem1.htm
http://www.stolaf.edu/people/giannini/flashanimat/lipids/membrane fluidity.swf
http://biology.clc.uc.edu/courses/bio104/lipids.htm
http://www.stolaf.edu/people/giannini/flashanimat/molgenetics/dna-rna2.swf
http://academic.pgcc.edu/~ssinex/blt/nucleotides/nucleotides.htm

Enzymes

http://highered.mcgraw-hill.com/sites/0072495855/student_view0/chapter2/animation__how_enzymes_work.html
http://www.stolaf.edu/people/giannini/flashanimat/enzymes/enzyme.swf
http://www.stolaf.edu/people/giannini/flashanimat/enzymes/transition%20state.swf
http://www.stolaf.edu/people/giannini/flashanimat/enzymes/chemical%20interaction.swf

Plant Cells

http://www.cellsalive.com/cells/cell_model.htm
http://www.forgefx.com/casestudies/prenticehall/ph/cells/cells.htm
http://www.biology4kids.com/files/cell_main.html

http://micro.magnet.fsu.edu/cells/
http://telstar.ote.cmu.edu/biology/MembranePage/index2.html
http://www.stolaf.edu/people/giannini/biological%20anamations.html (Be sure to note the spelling of "anamations" here.)
Plant Physiology, 5th ed., by Lincoln Taiz and Eduardo Geiger, Sinaur Associates, 2010, and its companion website: http://5e.plantphys.net/categories.php?t=t

Soil Testing Laboratories

Check local cooperative extension in the United States and agricultural agencies in other countries. Land grant and other universities often test soil as well, and these are not included in this list.

For a list of alternative soil testing laboratories in the United States provided by the National Sustainable Agriculture Information Service, see https://attra.ncat.org/attra-pub/viewhtml.php?id=285. In addition, the website of the Certified Organic Associations of British Columbia (http://www.certifiedorganic.bc.ca/rcbtoa/services/soil-testing-services.html) lists soil testing services for Canadian farmers, including both Canadian and U.S. laboratories. Soil Food Web affiliate laboratories in the United States, Canada, United Kingdom, Australia, and New Zealand are listed at http://www.soilfoodweb.com.

A&L Eastern Agricultural Laboratories, Inc., 7621 Whiteline Road, Richmond, VA 23237; (804) 743-9401; http://www.al-labs.com
A&L Western Laboratories, Inc., 1311 Woodland Avenue, Suite 1, Modesto, CA 95351; (209) 529-4080; http://www.al-labs-west.com
Agri-Analysis, Inc., 280 Newport Road, P.O. Box 483, Leola, PA 17540; (717) 656-9326; http://www.agrianalysis.com
Agri-Energy Resources, 21417 1950 E Street, Princeton, IL 61356; (815) 872-1190; www.agrienergy.net
Agronomic Division, North Carolina Department of Agriculture, P.O. Box 27647, Raleigh, NC 27611; (919) 733-2655; http://www.ncagr.gov/agronomi/sthome.htm
Analytical Laboratory and Maine Soil Testing Service, 5722 Deering Hall, Orono, ME 04469; (207) 581-3591; http://anlab.umesci.maine.edu
Brookside Laboratory, Inc., 308 South Main Street, New Knoxville, OH 45871; (419) 753-2448; http://www.blinc.com

Kinsey's Agricultural Services, 297 County Highway 357, Charleston, MO 63832; (573) 683-3880; http://www.kinseyag.com

Logan Labs, LLC, 620 North Main Street, Lakeview, OH 43331; (888) 494-7645; http://www.loganlabs.com

Midwestern Bio-Ag, 10955 Blackhawk Drive, Blue Mounds, WI 53517; (800) 327-6012; www.midwesternbioag.com

Midwest Laboratories, 13611 B Street, Omaha, NE 68144; (402) 334-7770; www.midwestlabs.com

Spectrum Analytic, Inc., 1087 Jamison Rd NW, Washington Court House, OH 43160; (800) 321-1562; http://www.spectrumanalytic.com

Woods End Research Laboratory, P.O. Box 297, Mt. Vernon, ME 04352; (207) 293-2457; www.woodsend.org

Organic Fertilizer Calculators and Worksheets

These calculators and worksheets help you determine what natural fertilizers to mix together to achieve desired nutrient levels.

Oregon State University: http://smallfarms.oregonstate.edu/node/175833/done?sid=10

University of Georgia: http://www.caes.uga.edu/publications/pubDetail.cfm?pk_id=7170#Fertilizer

How to Gather Soil Samples

http://njaes.rutgers.edu/soiltestinglab/howto.asp

http://soiltest.cfans.umn.edu/howtosam.htm

http://www.youtube.com/watch?v=qCbO5a5JZpk

Glossary

acid: a substance that produces hydrogen ions (H^+) in water

actin filaments: threads that serve as structural support for the cell, as well as part of its transportation infrastructure and communication network

active transport: using energy to move a molecule against its concentration gradient

adenosine triphosphate (ATP): energy currency of life; this molecule has two phosphate-to-phosphate bonds that when broken provide energy

adhesion: tendency of dissimilar molecules to attach to each other, as with water and glass

amino acid: molecular building block of proteins

anion: negatively charged molecule

apoplast: extracellular pathway created by the connection of plant cell walls

aquaporin: membrane protein that allows water molecules to pass through

atom: smallest building block of an element; consists of a nucleus surrounded by a cloud of electrons

base: a substance that produces hydroxyl ions (OH^-) in water

biofertilizer: fertilizer provided by living organisms

carbohydrate: carbon-based molecules consisting of combinations of carbon, oxygen, and hydrogen, where there is usually one water (H_2O) molecule for each carbon atom

Casparian strip: single layer of wax-clogged cells at the endodermis of a root

cation: positively charged molecule

cellulose: complex molecule made up of simpler glucose woven closely to form long strands

cell wall: strong, lattice-like structure that surrounds a plant cell; made up primarily of cellulose

channel proteins: tunnel-like, gated proteins that allow ions to move across membranes passively

chlorophyll: molecule that absorbs energy from light and is essential for photosynthesis; it gives plants their green color

chloroplasts: organelles that contain chlorophyll

chlorosis: inability of leaves to produce enough chlorophyll, resulting in yellow leaves surrounding greener veins

chromoplasts: cells that contain the red, yellow, and orange pigments found in flowers, fruits, and some roots

cohesion: tendency of the same type of molecules to stick to each other

collenchyma: thick-walled cells that provide support and flexibility to plants

compound: two or more molecules of different elements

cristae: compartments in mitochondria created by folding the inner membrane

cytoplasm: liquid and organelles (minus the nucleus) that make up a cell

cytosol: clear, jelly-like substance that is the major component of cytoplasm and literally holds the cell's organelles

dermal tissue: cells that form the skin of plants, consisting of the epidermis and the periderm

desmotubule: small tubular structure found inside each plasmodesma

diazotrophs: soil microbes responsible for nitrogen fixation

diffusion: passive movement of molecules from regions of high concentration to those of low concentration

disaccharide: molecule that consists of two glucose molecules linked together

DNA: deoxyribonucleic acid, a molecule that contains the genetic code

electron: a negatively charged particle

element: a substance that cannot be separated into simpler ones

endoplasmic reticulum: organelle that serves as the major pathway for the transport of cellular material

enzyme: biological molecules that increase the rate of chemical reactions

food web: series of food chains linked together

Golgi apparatus: organelle responsible for protein transport

ground tissue: cells that provide most of the mass or bulk of a plant as well as its support

heavy metal: metal with a high specific gravity that is usually toxic

humus: extremely stable organic matter that is resistant to further decay

hyaloplasm: clear, fluid portion of the cytoplasm

hydrogen bond: a bond that forms between a hydrogen atom and a negatively charged atom that is itself part of a molecule or a group of atoms

hydroxyl ion (OH^-): negatively charged molecule consisting of an oxygen atom and a hydrogen atom

hypha (pl. hyphae): thread-like filament formed by a fungus

integral membrane proteins: specialized proteins embedded in the plasmalemma that allow for transport through the membrane

ion: charged molecule

ionic bond: bond created by the mutual attraction of oppositely charged ions

ion pump: see protein pump

Law of Return: recycle all plant and animal waste to keep producing humus

Law of the Minimum: plant growth is limited by the least abundant mineral

leucoplasts: colorless cells located in parts of plants not exposed to light; used primarily for storage of lipids and proteins

lipids: organic molecule characterized by nonsolubility in water, making them hydrophobic; include fats, waxes, steroids, and triglycerides

lysosome: small organelles that contain hydrolytic enzymes that digest large molecules (mostly proteins) into their components; the cellular recycling centers

macronutrients: nutrients required in the largest quantities; includes carbon, hydrogen, oxygen, nitrogen, phosphorus, potassium, calcium, magnesium, and sulfur

meristematic tissue: undifferentiated plant cells that have the ability to become any type of cell

messenger RNA (mRNA): specialized RNA molecule that transcribes the DNA pattern in the nucleus and carries the pattern to the endoplasmic reticulum, where it is translated into protein molecules

micron: one millionth of a meter

micronutrients: nutrients needed in minute quantities; include iron, manganese, zinc, copper, molybdenum, boron, chlorine, and nickel

microtubule: specialized tubular structure for intracellular transport

mitochondria (sing. mitochondrion): organelles responsible for using the oxygen in sugar to produce energy

molecule: two or more atoms of different elements that are bonded together

monosaccharide: single glucose molecule

mycorrhizal fungus: fungus that forms symbiotic relationships with plants; brings nutrients and water to the roots in return for carbon-based exudates produced by the plant

nanometer: one billionth of a meter

neutron: subatomic particle found in the nucleus of the atom that does not have a charge

nitrogen fixation: breaking the strong bonds that hold nitrogen molecules together so that nitrogen becomes biologically available

nonpolar covalent bond: bond between atoms wherein electrons are shared equally

nucleic acid: a type of molecule that includes DNA and RNA

nucleotides: building blocks of nucleic acids

nucleus: organelle that contains the DNA of a cell

osmosis: diffusion of water across a semi-permeable membrane

oxidation: loss of electrons in a chemical reaction

parenchyma cell: specialized cells in which metabolic functions occur but that retain the ability to become further specialized

passive transport: movement of molecules without the input of energy

pericycle: outer boundary of the stele, consisting of a thin layer of cells that still have the ability to divide

peroxisome: digestive vessel that is involved in the conversion of fatty acids into sugars

pH: measure of hydrogen ions in solution; liquids with a high pH are basic or alkaline, and those with a low pH are acids

phloem: tissue responsible for the transport of nutrients throughout plants

phytoremediation: use of plants to remove or neutralize toxins

plasmalemma: double-layer phospholipid membrane that surrounds plant cells and is located just inside the cell wall

plasmodesmata (sing. plasmodesma): tunnels that connect individual plant cells

plastid: cellular mini-factories and storage facilities in plant cells that include chloroplasts, leucoplasts, and chromoplasts

polar covalent bond: a bond between atoms wherein electrons are not shared equally, thus forming a region in the molecule with a negative change and another with a positive charge

polysaccharide: chains of simple sugar molecules

protein: nitrogen-based molecules composed of amino acids, each of which has an amino group (NH_2) and a carboxyl group (COOH), plus a side chain whose composition can vary

protein pump (or ion pump): proteins that pump ions across a membrane against their energy gradient

proton: positively charged particle within the nucleus of an atom

reduction: addition of electrons in a chemical reaction

rhizosphere: area immediately surrounding roots

ribosomes: organelles where mRNA is translated and proteins are made

RNA: ribonucleic acid, a molecule used to pass genetic information from the nucleus to the ribosome

root hair: specialized epidermal cells that increase the surface area of roots

salt: a molecule that is the end product of a neutralization reaction, in which two ions neutralize each other's charge

schlerenchyma: cells that develop a lignified, secondary wall and can no longer elongate; these provide support and physical protection to plants

siderophore: chemicals produced by some rhizosphere microbes that free iron in the soil

statoliths: organelles in root cells that act like weights, sinking to the bottom in response to gravity

stele: central part of a root

stomata: openings in leaves that let in carbon dioxide and let out oxygen and water vapor

suberin: waxy substance that waterproofs cells

symplast: area encompassing the inside of all connected cells

thylakoid: flattened, hollow discs inside chloroplasts that contain the chlorophyll antennae that absorb energy from light

tonoplast: membrane that surrounds plant vacuoles

transfer RNA (tRNA): specialized RNA that carries amino acids to the ribosome during protein production

trichomes: hair-like structures that aid in the evaporation of water from leaves

vacuole: membrane-bound spaces inside plant cells

vascular tissue: combination of xylem and phloem

vesicle: membrane-bound structures that transport materials inside the cell

xylem: tissue responsible for the transport of water from the roots upward to the leaves

Index

PHOTO AND ILLUSTRATION CREDITS

Michael Amaranthus, Mycorrhizae.com, pages 136, 211
Jeremy Burgess, Science Photo Library, page 226
Dave Carlson Studios, Carlson-Art.com, pages 23, 30, 37 (right), 43, 45, 48, 50, 86, 91
Alessandro Catenazzi, page 207
Markus Dubach, U.S. Department of Agriculture, page 210
Steve Hillebrand, page 209
Judith Hoersting, pages 170, 187, 219
Tom Hoffman Graphic Design, pages 56, 57, 59, 65, 159
Louisa Howard, Dartmouth Electron Microscope Facility, pages 77, 81
Louisa Howard and Charles Daghlian, Dartmouth Electron Microscope Facility, page 82
Dennis Kunkel Microscopy, Inc., pages 2, 24, 25, 44, 74, 76, 79, 85
Kari Lounatmaa, Science Photo Library, pages 39, 41
Darius Malinowski, page 196
Arthur Mount, pages 101, 103, 128, 144, 182
Arthur Mount, based on illustration by Tom Hoffman, pages 14, 113
Pietro M. Motta and Tomonori Naguro, Photo Researchers, Inc., page 37 (left)
Markus Nolf, page 46
Gary Raham, Biostration.com, pages 27, 29, 35, 42, 78, 116, 124, 130
Hermann Schillers and Dr. H. Oberleithner, Science Photo Library, page 33
U.S. Department of Agriculture, Natural Resources Conservation Service, page 212
U.S. National Aeronautics and Space Administration, page 199
Keith Weller, page 87
Laken Wright, pages 142, 161